NIGHT LIFE

Also by Diana Kappel-Smith
Wintering

NIGHT LIFE

Nature from Dusk to Dawn

Diana Kappel-Smith

Illustrations by the author

LITTLE, BROWN AND COMPANY

BOSTON   TORONTO   LONDON

FIRST EDITION

Library of Congress Cataloging-in-Publication Data

Kappel-Smith, Diana.
    Night life : nature from dusk to dawn / Diana Kappel-Smith ;
illustrations by the author. — 1st ed.
        p.   cm.
    ISBN 0-316-48300-1
    1. Nocturnal animals — United States.   2. Natural history — United
States.   I. Title.
QL755.5.K36   1990
591.5 — dc20   89-35197
                                                                        CIP

10 9 8 7 6 5 4 3 2

Designed by Robert G. Lowe

FG

Published simultaneously in Canada
by Little, Brown & Company (Canada) Limited

PRINTED IN THE UNITED STATES OF AMERICA

For my mother,
Victoria

Acknowledgments

I owe an enormous debt of gratitude to the people who have made this book possible. Julie Fallowfield and Ray Roberts have kept their parts of a good bargain and have kept me to mine. Pete Dunne gave me confidence. Mary Main — a best-selling author and a friend, who has been, for the last forty years, blind — has made me aware of what the miracle of sight is. And of what it is not.

Many of the following people appear in this book, often on a first-name basis. Other people and organizations do not appear in person, but they have all smoothed my way and guided me, to people and places and sources, and to knowledge, and to experience. The only way I can think to pay my debt to them is to name them here.

G. Stuart Keith, Research Associate, Department of Ornithology, American Museum of Natural History; Dr. Wade Sherbrook, Resident Director of the Southwestern Research Station of the American Museum of Natural History; Dr. Howard Topoff, Professor of Psychology, Hunter College, SUNY; Dr. Jan Randall, Professor of Biology, San Francisco State University; Joseph R. Mendelson III; David Pfennig; Richard Taylor; Randy Morgan; William Kambitch; Mr. and Mrs. James P. Richards; The Amerind Foundation; Mary Erickson, of the Arizona-

Sonora Desert Museum; Dr. Stephen H. Allen, Furbearer Management Supervisor of the North Dakota Game and Fish Department; Bonnie and Brent Woodward; Mr. and Mrs. Gary Jepson; Calvin Cobb; Bill Austin; Daro Crandall; Leo and Shirley Olsen; Gene Kohler; Gary Billings; John Schumaier and his grandfather; John S. Strazanac; Aubrey Moore; Dr. Rosemary Gillespie, Professor of Biology at the University of the South; Dr. Kenneth Y. Kaneshiro, Director, Hawaiian Evolutionary Biology Program; William Walsh, PhD, President of the Coral Reef Foundation; Dr. Francis G. Howarth, entomologist, Bishop Museum; Dr. Arnold Suzumoto, ichthyologist, Bishop Museum; Dr. Minoru Tamashiro, Professor of Entomology, University of Hawaii; Marj Awai, PhD, aquarium scientist at the Waikiki Aquarium; Dr. Francisca C. do Val, Director of the Museu de Zoologia, São Paulo; Carol N. Hopper, PhD, Science Advisor, Education Department, Waikiki Aquarium; Sandra Romano; Colin J. Lau, PhD; Sam Gon III, PhD; Rob Rydell, of the Nature Conservancy of Hawaii; William Mull; Julio de la Torre; Dave Norris, science teacher, Joel Barlow High School; Dr. Charles Remington, Professor of Biology and Curator of Entomology, Yale University; Dr. Lawrence F. Gall, Curatorial Associate in Entomology, Peabody Museum of Natural History, Yale University; Richard Haley, naturalist, New Canaan Nature Center; John Schuhe; Johnnie W. Tarver, Chief of the Fur and Refuge Division of the State of Louisiana Department of Wildlife and Fisheries; James O. Nunez, Chief Refuge Warden of Rockefeller Refuge; Craig A. Guillory, Refuge Warden, Rockefeller Refuge; Junior Vaughn, Refuge Warden, Rockefeller Refuge; Malcomb Agbert; Bill and Bonnie Dekemel; Milt and "Red" Dudenheffer; Frank and Audrey and Mike Charpentier; Dave Hall, of the U.S. Fish and Wildlife Service; Allan Ensminger; John "Frosty" Anderson; Dave Taylor, PhD; Noel Kinter, PhD; Dr. Ted Joanen and Larry McNease, Louisiana Wildlife and Fisheries Commission; W. Guthrie Perry, fisheries biologist; and the owner of the oldest bar on Bourbon Street, whose name escaped me but whose words did not.

NIGHT LIFE

Now I think it is true to say of the road, and also of God, that it does not move. At the same time, it is everywhere. It has a language, but not one I know. It has a story, but I am in it. So are you. And to realize this is a moment of some sadness. When we are denied a story, a light goes off . . . do we vanish too? I am asking you to study the dark.

— Anne Carson, *Kinds of Water*

Introduction

I began this book, journey, same thing, with a simple desire: I wanted to know what wildlife was out there at night and what it was doing while we were — most of us — asleep; or, anyway, not paying attention. I use the word "wildlife" in the broadest sense. It turns out that night living is the majority view, practiced by most animals (and some plants) so that even in limiting myself to North America I saw right away that the scope of the journey was going to have to be wide, too wide.

I began with a kind of map. At first, there were three sheets of paper. On one was a list of five very different places to be found in the United States. On another was a list of creatures that were active at night; creatures that were nocturnal, in other words. Or crepuscular — a scabrous-sounding word meaning active-at-dawn-and-dusk. This list contained most major animal groups: spiders, insects, fish, amphibians, reptiles, birds, mammals of various orders, et cetera; it was very rigorous, a taxonomic structure, as though I were outlining a text. I threw people in for good measure, not because we're nocturnal by nature but because some people have become so, by profession. The third piece of paper held a list of technologies, or sensory arrays: smell, taste, hearing, vision, touch, and their derivatives: radar, infrared sensing, chemical communication, and so on. Seasons entered into the maps, too. Endless variations and interweavings of all of the above were, of course, possible.

So I took a pair of scissors and chopped the lists up into single words and put them in a bowl that had held scented herbs; the dry leaves and petals were dumped, making room for the verbal potpourri. The bowl sat on my desk for almost two years. As a place was visited, or a creature met, or technology explored, I took its word out and threw it away. The bowl is now empty.

It was a good map. It took me where I needed to go, farther than I had intended.

Please note that a writer is a kind of craftsman, like a carpenter; some carpenters are good at city halls and others do cabinets. I'm the kind who is most comfortable with wood-grain and closure of joints; I learned my craft on small-town newspapers and glossy magazines where the subjects were temporal and the point of view was close. I write most easily in bits, essays, chunks. This is a book of chunks. Sometimes a chunk contains a community of creatures — a coral reef or a piece of desert; others contain only a single kind of creature — a coyote, an octopus, a fruit fly, each intended in a reflective way as an ambassador of its kin. This is not a text, full of generalizations. This is about the particular. But remember the map: the map does encompass a whole, like a mosaic of bits that resolves, at a distance, into something larger than the single chips suggest.

This book is also a journey, and if some chaos has assembled itself around the central pursuit then I've left it in, because I've enjoyed it. Every journey has an element of beachcombery and this is no exception; I hope that you will enjoy the gritty oddments as much as I have.

For the most part, this has been written in the first person and the present tense, as though it were perpetually happening now. Some things happen only in the *now* and these were the important things. So be it.

One of the purposes of an introduction is to introduce myself; a sort of handshake-and-good-evening and a titillating one-liner by a gracious host: "This is the woman I told you about, the one who does the marvelous translations of Federico García Lorca . . ." Well, one day perhaps I will be introduced like that, and will sprout wings. To translate García Lorca is to attempt the impossible; I am convinced that it cannot be done, that it can only be attempted, but that is exactly its attraction.

I see that I've already shot off on a tangent, and that I ought to begin again.

So: I don't translate García Lorca, really, though I've experienced the ecstasies and frustrations of trying. I have become a translator of science and nature; a job that I am convinced can also never be done, only attempted. Science is our way of penetrating nature, and thus hovers always on a brink: it stands with its heels on a body of knowledge — a body that is always quaking with revisions — and its toes are poised over mystery. If the scientific method seems cold and overdisciplined at times, how else, I ask you, would you hold on? A scientist, like a sky diver, is careful, because the name of the game is to take that one step out.

As for Nature — with a capital N — this word has been washed by so much surf of magnificent declamation that I hesitate to add any more right here. This whole book is my addition, though it can't properly be called mine: it has been done with the knowledge, experience, and guidance of too many others to be so blithely owned.

It is now also owned by you, I presume. A book is a presumption: it asks for your time and attention, and gives some specks of data, and perhaps amusement. As I began with a bowl full of snipped words that grew into this, I hope that a word or two in here will be taken up by you and will grow into something new and unimaginable. That would be natural. That is what I hope.

ARIZONA   August

But she realized now that here on this planet there was no need for color, that the greys and browns merging into each other were not what the beasts knew, and that what she, herself, saw was only the smallest fraction of what the planet was really like. It was she who was limited by her senses, not the blind beasts, for they must have senses of which she could not even dream.

— Madeleine L'Engle, *A Wrinkle in Time*

Rains Far Away

From the air it does not look good. It is pale and creased and sprinkled with darkish speckles. I wonder why I am going.

I have a lot of answers ready for myself but I find that I am not convinced. They sounded OK back home in February and now they don't.

I know I'm coming here because this desert in late summer is so hot that most creatures are active at night. I've read books about this and I've talked to people who know. I've made lists: owls, bats, toads, ants, snakes, kangaroo rats, coatis, mountain lions. This is what I want to see and Arizona is the place to go. The summer monsoon rains start in July or early August and all kinds of animals come out of the woodwork.

It looks exciting, on paper anyway. All Systems Go for Arizona in August.

Months ago I made arrangements to stay in the mountains near the New Mexico border, at the Southwestern Research Station of the American Museum of Natural History. This sounds official and honorable and fine and I've called and they're expecting me and I have a map and directions, and I have boarded the right plane; maybe it's the light that's wrong? I don't know.

The light is diffuse and seems to come from nowhere and lays no hard edges anywhere. The ground is the un-color of dust and the vegetation is like greenish gray pills of fluff rubbed up from a worn rug. From here, the desert looks like a lot of nothing. I do have all my files of photocopied notes and correspondence that say otherwise, and my photography gear and field guides and night glasses and packs and sharpened pencils and rechargeable flashlights, but it seems frail stuff with which to tackle that.

We are over Tucson now and coming down. The city looks thinly spread over the nothing like a mosaic inlaid with flecks of swimming-pool turquoise. Everything has the same dun grouting of dust. I can see cars crawling, candy-colored metallic lozenges, squared-off ants. There is the tinselly glitter of a trash dump and a mottled patch of lawn like water-stained fabric. A stewardess announces that it is 115 degrees Fahrenheit down there. I know as a certainty that I am not going to like Tucson.

And I don't; but that's neither here nor there. There have been cities that I have loved and cities that I have endured but never one that has given me quite this sense of vacuum. Once I am down and out, I plod from ovening heat to canned cool and back and forth again and manage to board the right buses and navigate to a rent-a-car, a post office, and a Quick Mart. With Diet Coke between my knees and map spread on the starboard seat I am on my way.

I am going southeast toward the New Mexico border and I have to find a canyon there. That's all I know.

The journey is long and the air-conditioning in my pint-sized auto isn't equal to the job. The air tastes fused. Light seems to come not so much from the sun as from the ground, from the panlike plains of scorched and raddled grit. I dodge from one exit oasis to the next, consuming Diet Cokes and perspiring mightily in between, and the appearance of nothingness goes on and on. The desert doesn't get interesting until I get close to the mountains. Things are happening in the mountains.

Before I get into the mountains I go through a breadth of country through which little fences have been stitched, and there is a road, or two, or three, always running straight. The roads look stuck down like lengths of gray tape. On the low ground there are fields of alfalfa that are green and thick-napped but that also look pasted on, held there by the silvery pipes of their irrigation machinery. The ranch houses are as substantial as toys. It occurs to me that the desert is a big place and that the twentieth century looks silly in it.

The distant mountains are old and worn and darkish like bad teeth, gullied and wrinkled and collapsed like dried-out sand castles. Between the mountains the land is flat but the mountains have crumbled down around the edges of the plains and have rounded them like bowls.

The scale of all this basin range is very large, but because the air is so dry and clear I can see a long way and none of it looks large. When I crawl over the rim of one basin into the next, the far edge seems to be right over there and only after the car has been groaning away over that basin floor for half an hour do I get a sense of the size.

The vegetation is so scanty that I can see the pale skin of the land everywhere, and here and there bone; buttes and rocky knobs and — at the rim of a pass, in the yawn of a canyon negotiated between one basin and the next — tumbled rounded boulders so enormous that I find myself driving, hunched, on the centerline of the road. I look at the boulders out of the corners of my eyes, expecting them to roll like giant fists and crush me flat.

After I've crossed the pass I see that one distant range has clouds over it so that the mountains are black. I find this refreshing. Anything is better than the endless colorless ground with its scribbles of parched bush. Rents in the cloud let down light so that here a slope and there a peak is washed with a gold as friable as powder. The cloud is inhabited by pulsing white light. The country music on the radio is fried by static and I turn it off. I open the window and the air that jams in is as hot as a blast furnace and smells of wetted dust.

I close on the mountains slowly, and I keep the window open because of that smell, which seems exciting to me though I can't, at first, say why. Perhaps it's because of the promise of water that it holds out. Even in my hermetically sealed Toyota I've come to an understanding of thirst.

An hour later the mountains ahead of me don't seem to have come any closer, but they are rising up. I am moving southwest now, the road skirting the foothills. The light has changed again. The clouds have risen and are shutting in the light, or perhaps it is evening now; I have lost all sense of time. Though my watch says a number it makes no sense in the here and now. I have come three hours from home and across two thirds of a continent, and have been so dislocated by the landscape that what counts here now is light, not time.

Now I see that there are mountains farther west and that they are also rising and that they are a violet black. I know those must be the Chiricahuas and that the canyon where I am going will be in there. I have been warned not to arrive in the dark because of the difficulty of finding the way.

Aside from the odd enervated cow or steer I have seen no life on this desert, animal or human. The clouds are more active than anything else, they have all my attention; the storm in them is violent enough. I am crawling into the foothills, into the storm. These foothills are part of the line of mountains that I will come to know as the Peloncillos; they will form the far rim of my home basin for the next two weeks, but I am among them now. I crawl through them slowly. Though the speedometer reads 70 all the time I seem to be caught in one of those nightmares in which one's feet are nailed to the floor.

Lightning judders in bursts and everything trembles, even me. I keep stopping and getting out to look and the wind whips my skirt over my head, plasters my hair across my nose; the bristly chias that line the tarmac, burry and sand-dry, claw at my ankles and draw blood.

There is rain falling but it does not fall here, it falls a few hundred yards off (it may be miles off; I'm not sure anymore and have learned already not to trust my sense of distance) but there it does fall. And the rain is orange. Orange rain! And there is a rainbow, naturally, which lands in misty light on the flank of a soft little hill. I have to restrain myself from running off after it, this rainbow-end that looks so strong and so close.

In the hours ahead there will be the slow climb up the basin rise and into the canyon, and the winding way up through towering buttes and deepening forest and already deep dusk, and the getting lost and the turning back a dozen times before coming to the Research Station and my room and bed. Still ahead of me is the shock when I open my duffels of finding the things in them just as I packed them — my green toothbrush, my hiking boots a little squashed, my old T-shirt only wrinkled and warm from a journey that seems to me to have been infinitely long. The wrinkles will shake out and the fabric will cool as I put it on and know that I am hungry, and tired.

I will find in my first days in this desert that it always rains far away. In the afternoons the clear sky will sprout puffs of cloud that will gather and darken; suddenly, across the basin floor, paved with its dust and puffs of mesquite, thunder will roll as if the canyon were a drum and lightning will rip like a magnesium flare. Rain will go down in the distance like a curtain dropped, a column of grayness streaky but oddly static, like a scrubbing of pencil, regretted, and badly erased.

I will see this and I will wait for the rain to come with impatience and with a kind of hunger new to me. I will think of the rain as an animate thing; powerful, fickle.

Now, here, there is only this rainbow, which is creeping softly through the foothills and which has a nether arc, so that there is a whole bow of color against orange rain and black cloud. It curves ahead of me like a proscenium arch. The little car noses bravely along its tape of road and I am getting there. This is all that matters: that night is coming onstage so softly, a cool cloak made of mountain shadow and cloud and a whole planet's turning away. It is a good darkness. I am sure now that it will be full of life.

Territory

Lust for territory is as old as the hills. The hills that I've come to live in here are made mostly of rock that was laid down in oceans, and the oceans were full of living things, and living things have territories. Even limpets crawl home before high tide to the hollows that they've quarried in the rock. Corals wall invaders off and anything unwise enough to scale their battlements will be met with a vicious sting. Even a thing as humble as a penicillium mold puts out a chemical that discourages other molds, and bacteria, too, which is why it works in us. The walnut trees in the canyon let a mild herbicide seep from their roots to keep other plants from growing where they do. They stake their claim.

It's natural for me to want to stake mine. If I want to know this desert it makes sense to get on intimate terms with some small piece of the thing. Desert is complex; no good definition exists. One can say that deserts are dry, but dry in comparison to what? Dry always and everywhere? How dry?

I looked all this up in books back home. I looked "Desert" up in the *Britannica*.

"Generalizations," said *Britannica,* "are poorly advised."

I read on. *Britannica* is on its high horse. On the subject of deserts it sounds like a scholarly uncle who can't quite trust you to appreciate the facts; a bachelor scholarly uncle (any wife worth her salt would have trounced this stuffiness long ago) but I keep pleasant company with him anyhow, overfond as he is of vocabularial show-off: "Both physiographic and geological variations abound . . . and even the ages of deserts are non-uniform. . . ."

I've consulted other books and absorbed more. I know that this is Chihuahuan desert, and not Mojave or Sonoran; but I know that some mingling of Chihuahuan and Sonoran is likely to happen here, and interminglings with mountain flora and fauna, too, so I am ready for anything.

Variations abound: basin and foothill, south face and north

face, steep slope and gentle. There is ground that is rocky, gravelly, sandy. There is land that is grazed and ungrazed by cattle. Each slope, each exposure, every variant of ground is home to its own community.

I want my territory to include as many of these communities as possible. But it can't be too big, because intimacy is what I want. And since I'm going to be there at night it makes good sense to know it well enough so that I don't get lost.

This is a tall order. I am full of trepidations as I set off in my little car. My hiking boots are on and so is my hat, I have a map and another six-pack of Diet Cokes nestled in a two-dollar cooler in the trunk, and I don't think I'll find the place.

By noon on that first day I have found it. It's a canyon that runs east-west so that its slopes are of opposite kinds; the north-facing rise is scrub desert, scattered mesquite bushes mostly — the darkish speckles and dusty pills of yesterday. There are patches of grass between them and dry evidence of wildflowers; otherwise gravel and dust. The south-facing slope is steep and rocky and full of prickly pear and clumps of agave and sotol; a classic patch of succulent desert. Between the two is a wash bordered by real trees — juniper, sycamore, live oak, black walnut. These are the broad botanical outlines, and they are fine. I have called the rancher who has grazing rights there and he will allow me to go whenever I wish as long as I don't "booger the cows."

I assure him that I won't and I wish I could assure myself that they won't either. I don't trust cows, but cows are easy enough to see and hear and are daytime creatures anyway, so I was confident that we wouldn't booger each other overmuch. It was the other things that the rancher said:

"Don't fall into those mine shafts now," he said, "they ain't fenced too good."

"Mine shafts?"

"Yep. Some fellers pecked around up in there years ago. Lived up there like coyotes. Never come up with anything that would buy beans," he said. "They left some holes in the country."

"Oh?"

"Uh-huh. Yes, ma'am," he said, "and it's fuller of snakes up there than a dog of fleas."

"Oh." .

There was a pause, then:

"Keep a sharp lookout now," he said.

"Thanks," I said.

"You're welcome to it," he said.

"'Bye now," we both said.

There it was. I'd staked my claim.

I found the mines later on, but what I'd found already was the water. The canyon has a spring. I'd found that right away. It doesn't look like much; a thin trickle seeps from the rock and wanders through stones for a few hundred yards before it disappears. Some of this is piped into a tank for the cows.

When I discovered the place, it was still early morning. I walked the wash below the spring and sat in the shade of a juniper, and fixed the jutting root in my mind for future reference. I heard footsteps, a crush of gravel. I froze. Three buck mule deer came down the wash, buds of horns in velvet, coats goldy-gray on lean flanks, big ears up and mobile. The wind was from them to me and I could smell their musk. They drank from the dark trickle and went on.

The name of the place is Silver Spring. That has a good sound, whether the riches are of water or elusive ore. At dusk I am back there with my pack of gear. I descend from the desert road into the little canyon as if I were entering a pool, a swatch of forest tucked like caulk in a crevice of ground.

Here a dozen ancient sycamores hold, literally, sway. They are spaced on the level ground, their whorled and bifurcated boles so large it would take five of me to circle one of them. In the pale gravel of the wash below them runs the streak of water no broader than my hands held side by side. This is their thread of life, too.

I take my pack off and lean us both against a sycamore to witness here, for the first time, the changing of the guard.

The breeze has dropped, the leaves stand still. Acorn woodpeckers do their evening bob-and-call along a branch and go to their night holes in the sycamores. Mourning doves slide down the shadowed wash; they call twice and are silent. Cassin's kingbirds serenaded my descent and still buzz from the desert slopes. Already there are bats flittering among the leaves.

The surface of the cattle tank reflects, in silver and verdigris, what there is here of sky. I hear a great horned owl — though I don't believe my ears at first. This is a familiar voice; his challenge and his territorial claim, the same deep commandment of hooo-hooos that I know from home. So I am amazed. I hold my breath.

So much of a naturalist's natural fix seems to come from the hunting and finding of the rare, even of the ridiculously rare. This is the place to find the rare. These mountains here have heaved up a biotic stew; I can trundle my Toyota from spruce-fir peaks to desert flats in forty minutes, and thousands of living things all seek their natural niches along the way. The birdlife here is the stuff of legend. Mexican birds drift up across the border into these mountains: Lucifer hummingbirds, ferruginous pygmy-owls, elegant trogons, dozens more. Birders from all over the world come to stay a night or two at the Research Station before steaming off to distant canyons full of fresh delights. I have absorbed enough new bird names and bird sights in my first day to make me dizzy. What I do not expect here is the commonplace.

The great horned calls again. I answer him in his own language. There is a pause.

This is a challenge to his turf. Great horneds have varied turfs. They are at home from the Pacific to the Atlantic, from Point Barrow, Alaska, through Baja California and deep into Mexico.

This is a big bird and a big success story, too, as birds go. This too is his home ground. I've challenged him one to one and now I wait.

Still nothing, no reply. A pause for contemplation? Is he fooled? Has he been played so many horned-owl tapes by birders that he's fed up? Or frightened off (this does happen)? No; he answers. The same hoo-hoos but from closer now. He's coming. I allow an owlish time to pass and hoot again.

Another pause. I've got him wondering. Or does he know I'm me now and has he shut up in disgust? Does he think I'm neighbor owl from upcanyon threatening his ground? (This is what I want him to think. I want him to come and see if I'm not just another binoculared imposter. I want to see him.)

I do: here he comes, a bulky silent float through leaves. He grasps a branch, flips his wings like a man adjusting the tails of a morning coat, tilts forward. And I sing again through my muffling fist. "Hoo-hoo . . . hi!" I say. "It's just me," I say, low, softly.

He doesn't believe me, quite. He waits, peering down. I look back, I can see him well enough; gray and tawny, heavy, ears perked, eyes fiercely browed. Huge eyes, all the better to gather scant light. He is a Presence here. Perhaps I need my own song?

I sing the first thing that comes to mind: "Go down . . . Moses . . . 'way down in Egypt laaaand . . ." My voice quavers at first but firms as I go on. He sits up. "Tell old . . . Pharaoh . . . to *let* my people go-o-o. . . ."

He waits through the verse about the firstborn before he nods and flies away. Then he's off upslope in a thick silent drift, bearclaw-sized talons furled beneath him. He's off to hunt — anything that moves. Desert cottontails that scutter over the desert road, young antelope jackrabbits, Gambel's quail, rattlesnakes . . .

I should not have thought about rattlesnakes. A mistake. It's dark now. I sit frozen to my sycamore trunk, and I keep myself looking up. I try another owl call. A call of a night bird, not a dusk-dawn bird like the great horned. I've been told to expect them here; the elf owls, true desert owls, minuscule puffs of bird whose calls sound like a bad imitation of a chimpanzee . . .

I try it out. I leave a decent pause — nothing. I try it out again. Bats flitter. There's a low drone of a passing beetle, an insect chorus of rattles and shrills like background static. Then: the hysterical titter of . . . yes. An elf owl. Yes. Very soft, very high overhead; the size of a sparrow; the hunter of moths.

Later: I have crunched and tripped my way down the gravel wash and have taken up residence on my juniper root. I scan the wash with my binoculars and drink some water. It is almost midnight and there is no moon. I am at the limit of my vision and my eyes tell me things that are not so: the shadows between the stones writhe like snakes and the binoculars assure me that they are shadows between stones, and without the binoculars they writhe again.

I don't want to turn on my light. If I do that, whatever night vision I have will be gone. So I sit.

I tuck my feet under the root. I sit still, converted to stone so that no local demons will take an interest in me. I know I'm scared, not of what I know but of what I do not know; of what I am not equipped to know except in the half-way of assumption, tattletale, implication, superstition, guess.

I have my field guides and I can goggle through my night-glass binoculars but I am afloat — only my rear end and the toes of my boots are in touch with anything at all. The owl can see the glint off my binoculars and hear the "Go down . . . Moses . . ." in stereo but I am mostly blind now and, in comparison with him, in comparison with the skunk who can hear the scrape of a worm's bristles in the earth, I am deaf.

Rattlesnakes. A rattlesnake can feel the vibration of a mouse's skip. His belly is down there in touch with it all. His lapping scales and long muscle and tines of ribcage hum the familiar song of *mouse* to him like plucked strings. Through the pits below his eyes he can feel the heat given off by a grasshopper mouse ten feet away, just as I could feel the heat of a campfire if I held out my hands.

And jackrabbits: the antelope jack's excessive ears tell him all about a coyote's fur brushing against an agave stalk but bring him no further news at all, though he waits there, for the ting of a loose stone, perhaps; but he hears nothing. The coyote's nose tells her all about the jack who came through here just now and nibbled on mesquite pods. She can smell the sap of the mesquite and the fresh urine mark that says *jackrabbit* and more than that; it says that this is the old buck she's hunted before and always missed. She hunkers and creeps, each pad testing each stone before she sets it down. She will not miss him now.

What am I missing? Everything. I have too little of sensory input to process any of it well. My eyes aren't made to make sense of moonless shadow and I lack the keen equipment of the natives. My senses are inadequate for this task. So my brain begins to tell me fibs. It begins to tell me *writhe rattle gape creep crunch* and other things, all unpleasant in the extreme. What I feel is *fear* — fear as potent as nausea. I could, now, all by myself, construct an entire dark mythology, a whole biota of evil, a —

There is a crash in the undergrowth and I can only sit. Three headless shapes clatter toward me down the wash. I have enough wits left somewhere to know that the wind flows from them to me so they won't know I'm here maybe but my nose goes into high gear and I can smell them. They smell fusty, skunky, sour, very bad. They turn toward the spring and I hear sucking, splashing. I catch the glint of an eye where no eye should be. They smell me at last and swing toward me, noses high, then they skip, shy off, trundle snortling up the bank and disappear in the shadow of a live oak. A mother peccary and two half-grown young. Good God. Did I smell as mysterious, as sour, as demonic to them as they did to me?

Maybe I'm not so ill-equipped here after all. I'm new to the business, though. New to the territory.

That's it: new to the territory. If I were home now, even in the dark, I might stub my toe or knock over the alarm clock but I could put my hand out for the bedpost and turn toward the door — even in the dark of a house, which is darker than any out-of-doors — because I know my territory. I know every blink, hum, shadow; every smell of cold coffee or soap; every sashcord and sofa pillow. I know these things without thinking.

As the nights pass, I begin to know this place, too. I know where the track rises and dips as it climbs from the canyon toward the old mines, and where there is a way across a hollow and up through agaves to a knoll where I can see the line of mountains. I know where the ravines are, the clumps of juniper, the nests of nocturnal harvester ants who plug their burrows at dawn and unplug them at dusk. I know the sycamore hole where the elf owl sleeps, and I know which way the peccaries will come, up or down the canyon, depending on the wind. I know where the coyote den is below the knoll. I know the way uphill, around the mine shafts and the worst of the cactus clumps, and higher still to a bare flank of rock where I watch the sun rise and set.

Between seven-thirty and eight in the evening it is twilight there, marked by the odd *whit* of a bird, distant cricket chirps. The western sky is still light and strangely opaque. One star is exactly overhead. Rushes of breeze sigh through the canyon and the ocotillos rattle; the evening breeze is cool, dry. The great horned calls from the sycamores. I watch him fly upslope and off. The plumes of sotol wave over me as if they were sentient, as if they were painting away light from the sky.

I am a dawn/dusk creature now, like so much else that is here: the hares, the cottontails, the deer. I sleep in the heat of the day and am here from dusk to midnight always, sometimes until past dawn.

The light of day is oppressive to me and I walk in it hunched and with my head down, my shoulders and my every movement crushed by the desert sun as if it were a weight. Now I am pleased with the strewn stars, the night sky clear and spangled as if a lid had come off a box.

At nine o'clock it is night. A moth lands in my lap, a cricket starts up at my feet. There is creaking in the hills as the stones cool.

The stones have stories, too, and I know a few of them. The rocks of this high slope are full of fossils: bivalves that look like clams and scallops, and crinoids with their stems like bead strings topped with lacy flower-feet; and horn corals, and cherty clumps and fragments of other things, too, that I do not know.

They look only a little dirty after their long interment, as if they could be washed off and go back to work. I have pried a few from the gray mother rock. My pocket is full of the little beads of the crinoid stems, flattened and round with a central hole like Oriental coins.

They bear no head of Caesar Augustus, no *E Pluribus Unum* eagle. They won't buy much; this common currency of the lower Mississippian, a kingdom long superseded by others no richer perhaps but more in step with modern times. The crinoid signature still has value, at least to me; the stone rings jingle in my pockets, they speak in their coinlike way of a time when Arizona was equatorial, when this was tropical tidal flats and rolling lowlands — 400 million years ago.

This was their territory then. They were buried alive here in some gout of silt washing off the land, some tilt or slip of sea-muck, the churn of a storm. Here they are still.

The rocks say more. Their pages are crumpled by upheaval, by continents adrift, by fire. After the scallops and crinoids had been buried for 200 million years or so, volcanoes arrived in Arizona. Monstrous bubbles of molten stone pushed at the old rock from underneath, melting some, vaporizing more. The parent rock shattered, tilted. This whole slope, seen from the eastern end, looks like a book of sedimentary pages set on its spine. Hot mineral foams traveled up through cracks like bubbles rising in a Coke can. They crystallized in clots of azurite and calcite crystal, and veins of copper and silver ore.

Time passed then no faster than it does now; day and night were the same all through the births and deaths of volcanoes, through the droughts of the interglacials and the long cool rains of glacial advance. Then, eleven thousand years ago, more or less, people came to hunt mammoths here. They used spear-throwers and leaflike Clovis points chipped from the fossil cherts and volcanic glass.

Five thousand years later there were farmers living in villages in these canyons. They planted corn and beans and squash. They hunted antelope and bison with bows and arrows. They made pots decorated with red and black and white glazes, and with the images of birds.

It is ten P.M. now and dark and time to go down. As soon as I get up and set one foot on the slope my fear of snakes assaults me again like a wave, like an ague, a fever; my terror of rattlers is out of my control. I swallow to suppress the giggles and stamp to solidify my knees; it's like a sickness. I've tried to get a handle on this in several ways, and one of these ways has been to 'fess up to one of the naturalists that I've come to know. He camps in South Fork canyon in a tent, and goes everywhere by bicycle or on foot. He's been coming here for years. I thought, maybe, he'd know the antidote.

He didn't. When I told him about my snake-shakes — my voice sounded tinny in my ears when I did this, I was sure I was going to be accused of some purely feminine Freudian malaise — he coughed, laughed, choked, and confessed to a similar, uncontrollable fear, of bears. He isn't concerned about snakes. I'm not concerned about bears. If something goes bump in the night I don't think bear. He does.

We talked about this. We came to the conclusion that fear of night is natural to our species, and naturally projects itself on some animal; it becomes identifiable, merely demonic, a convenient excuse. He's picked bears. Some people pick bats. Other people pick spiders. I've got rattlers. Like most analyses of sub-conscious moil, this only makes me feel ridiculous.

I swallow again, stamp, take a deep breath, and start down. I have snake-guards on my legs but the slope is steep enough that going up or down is as much climb as walk, and I have to keep at least one hand on the rock above me. This is not good. I've tried to ward my demon off by learning all I can about snakes, which are only animals after all, and the knowledge helps a bit but not enough. I know that rattlers won't run off when they feel me coming. What they'll do is wait. If I put my hand in the crevice or ledge where they are, they'll strike what they can reach; hands, shoulders, neck. They use their rattles as a warning sign, a Keep Off Me like a skunk's loud stripes. They don't rattle at hunted game. They rattle when they're threatened. They don't always use it then, either; each snake is heir to his own stubborn individuality. The uniqueness of all live things is a good thought to think of and generally a fine thing, but it makes rattlers unpredictable.

I've gone less than half a mile when I realize I've gone wrong. I've traversed too far, the slope is too steep, the loom below is a juniper I don't recognize. I stop and take stock. The knoll is down and to the east and between me and it are the mines which ain't fenced too good but there's no other way.

I fish my light from my pack and turn it on; a miner's head-lamp with a battery pack that hangs from my belt. I don't like to do this. As soon as it's on I see my immediate surroundings all right, but everything else goes; I've lost the knoll, the loom of the mountains across the canyon, the canyon itself. I see only a prickly pear, a hump of stone with an embedded fossil scallop, the fallen stalk of an agave. I move slowly down, crabwise, sid-ling between each cactus clump and rocky loom. Freudian anal-ysis and naturalists' curiosity have limited power; each shadow-shift here still has the sliding-stone color of a diamondback.

*   *   *

When I come to the knoll, I spend some time with the ants. They are *Novomessor cockerelli* — nocturnal harvesters — and watching them calms me down. It always does. I fish a peanut-butter cracker from my pack and crunch half up for them and eat the rest. Ants cluster at the crumbs and begin to haul them off.

A coppery beetle blunders into my light and falls on its back and is overwhelmed before it can right itself. The attackers swarm in for the assault, their jaws slither off the beetle's broad carapace but they cling to anything they can hold; the legs are severed and carried off. A gray moth blunders and falls; in minutes it is sawn in two and its wings are lopped and all the pieces have disappeared into the maw of the nest. Two male army ants, with plastic-wrap wings and long abdomens like bronze sausages, tumble to the same fate. But the beetle is a tough customer. An hour later he is still alive, his ruby eyes still shine up at me from the horde of his attackers.

The ants aren't busy just with this benison that falls from my headlamp. There's more to do. Lines of foraging ants snake off into the brush. Some carry back the long-awned seeds of desert grass, others insect wings.

What looks like a milling chaos is not. It resolves itself, gradually, into order. Organization here is as tight as Law. There are the foragers who trail off in search of food, the midden workers who clean the nest and arrange the trash pile that sifts below it like mine tailings. Maintenance workers are underground, building galleries, caring for eggs and larvae and queen, arranging storerooms and dispensing food. The inspectors are up here, the watch ants who keep the mouth of the nest clipped free of vegetation and recruit the workers to attack.

All in the dark. All so neatly arranged! All done in a language of chemicals: trail chemicals, recruitment chemicals, sex chemicals, alarm chemicals, rank and role chemicals. They use chemistry and touch as we use sight and hearing, as the means of communicating with each other and of knowing their world.

It isn't the same world. It couldn't be. There is nothing we obey as they do. Only the laws of good and evil; though even here we are given choice. They are given none. They obey without question, without doubt. They heave themselves at the prone beetle. They explore my finger, antennae tapping along, and one fixes herself to a fingernail and hangs on. I pick her off and return her to the group that eddies around my crumbs. She scoots in tight alarmed circles, jaws wide and up, and others leave the crumbs and join her wild mill, ready to latch themselves onto any threat, even me.

Perhaps they know, perhaps they don't know, that inches uphill from the mouth of their nest is a Carolina wolf spider, a velvety night hunter like a mini-tarantula; her body an inch long and ashy gray.

Perhaps they know — I think they do — the exact whereabouts of the mustard-yellow crab spider who has wrapped one of their workers in a silky pot and is sipping his juices not a foot downhill from my scattered crumbs. They walk all around her and leave her alone.

Making sense of these ants is like tracing fossils to a source in time and space, like following a vein of azurite into the mountains' roots, like building a vanished culture up from shards, chips, ash. We inhabit this same desert slope, the ants and I, but we do not share the same truth.

Just before midnight I go down to the spring and light a fire where there have been fires before. I eat a sandwich and drink water and watch the fire and the stars. At two A.M. bumps of thunder and flashes of light begin to the north, among the Peloncillos beyond the plain. This goes on for hours.

I think of the Apaches who live north and west of here and still speak their own language. This was their place once. Just over a century ago and just a few miles from here, Geronimo surrendered the Apache claim.

The Apaches believe in the *Giver of Life*. Long years ago the *Giver* saw that his people were living in a wrong way and sent the *Gans,* the mountain spirits, to teach people how to live properly and in harmony with the world. The *Gans* came to the Apaches, and when they left they put pictures in the rocks, graven pictographs, to remind the people of the teachings.

Around a fire at night the Apaches would hold a spirit dance, as the *Gans* had taught them to do. They sewed racks of lath to black cloth hoods and painted themselves as spirits. The racks held painted stars to frighten away evil powers, and white dots that meant *pure heart*. Clowns would dance with the hooded spirits — silliness being a godlike attribute. The spirit dances drove out evil and let harmony in. They brought balance into the universe.

I have no dancers around this fire. The notion of harmony among my people is a matter of private conviction, not of public accord.

I hear the yap and yammer of coyotes from the northern slope. A skunk comes down the wash, and drinks, noisily. I watch the peccaries come and go. Just before dawn the three mule deer arrive, and the hills go gray, and rose, and a rufous hummingbird drones and chirps along the wash. He passes over my head five times and zooms on. He is going up to claim his blooming nectar-filled agave for a day, two days, three. It is his way station on the route to Mexico.

It is dawn now, and the brown towhee comes to the live oak. She plants her greedily tweeting youngster under a mesquite and goes hopping to find beetles and bash them; she delivers them to her little one well softened up. I wait until she has caught and served up three beetles, then I put my fire out and go. The sun has not risen yet but the night is gone.

This will be my last night here. The rains will come this afternoon. I have other places to see, other lives to bear witness to, but there will be no other place that I will call mine, not like this. Aside from mesquite wood I have consumed nothing; aside from cracker crumbs I have distributed no largesse; but Silver Spring — and its dancers, and the ghosts of its dancers — have made their claim on me.

Boogie and Roll

We were all out by the picnic tables making tie-dyed T-shirts. Some of the research assistants had been to Douglas and had bought cheap white T-shirts and RIT dye and a couple of packages of rubber bands. We were having a good time.

We needed this. Perhaps it was just a distraction from the waiting, because waiting for rain can be oppressive, and a rainless desert is oppressive. It was a break from field work, too, the daily drudgery in which most of the people here had been involved for so many months. Maybe it felt so good because most of us had been going it alone, whatever *it* was, and this wasn't.

It felt like summer camp. Edith and Marie were making tie-dyed socks to match their T-shirts and one guy was making a pair of tie-dyed Fruit-of-the-Looms.

It made me think of rain dances. This canyon must have seen several millennias' worth of those, and they must have had this atmosphere, too, at least in the preparation stage; a compound of Halloween and Advent, of make-up and dress-up and creative splash. And there would have been a spiritual component for some, though most people must have been spectators more than anything, participants in that they enjoyed the fuss and feathers, the break from daily routine, the party.

Whatever the rain dance was to dry ground farming, our wrapping of rubber bands and hovering over dye kettles was a leaven that made science bearable. By science I mean the field work part of it, which is dogged, grim, and painstaking. Like farming, it can fail.

I don't mean that field work can't be exciting, only that most of it isn't. After a week at the station I'd absorbed this: that field workers have to be humble, inspired, stubborn. Edith, for example. She spent her days climbing into the high country to count pine seedlings. Apache and pinyon pines share the mountain slopes, and her experiment was designed to discover how well their seedlings survived in shade or in the open, in mixed stands or alone, and at varying altitudes. For weeks now she had been lugging water for miles up the mountain to irrigate the experiment's plots. If she didn't do this, a year's worth of research would shrivel and die. Most of the seedlings were dying anyway and she didn't know why: drought, sun, wrong soil, lack of the proper mycorrhizae, fungus disease; no one knew.

Marie had come from France to study and collect tarantulas, millipedes, centipedes, and vinegaroons. She had several terraria full of them already, in the animal room, but she wanted more. She fed the centipedes with apples and they seemed to like them; they ate holes through the apples and draped themselves gracefully in the holes like lanky fashion models posing in Pompeii. What she fed the other captives I could never see, if she fed them anything. These creatures are famous for doing without, as long as they're kept cool. All of them are nocturnal, and all of them are most active during the summer monsoon, so Marie and her quarry were waiting for downpour, too.

Dave had come from England to study the behavior of a particular spadefoot toad parasite. He'd gathered some toads in Douglas where the rains had come the week before, but he hadn't collected any parasite data at all. He needed mating toads for that and the toads were waiting for rain. Meanwhile the toads from Douglas were in the animal room in boxes of potting soil. He was going to ship them back to the lab in London just like that. They traveled in potting soil just fine and would live buried in it for months and months. Whenever he wanted to get any toads up out of the potting soil he would put an electric fan on top of their box and turn the fan on, and the toads would pop to the surface; the rattle and bump of the fan overhead would sound, to the toads, like desert rain.

The monsoon was what we were waiting for; each of us was waiting for all our separate and particular reasons, but we weren't worrying about it much that afternoon. There were fifteen or twenty people out there clustered around the picnic tables and the kettles of green and yellow and red and purple dye, when it began to rain.

There was whooping and squealing and grabbing of kettles and rubber bands and we ran around the corner into the laundry room. It was noisy and steamy in there and some people brought out sodas and beers and started passing them around. The finished T-shirts were hanging up to dry overhead like carnival flags.

Somebody mentioned toads. Some people stood up then and looked around. Maybe this was *the rain,* but we weren't sure. Maybe it would peter out or sink into gravel and disappear before it reached the desert. Anyway, there was going to be a party tomorrow night and we were all going to wear our T-shirts. Some of the T-shirts were long enough to be dresses.

At dinnertime it was still raining so hard that you couldn't hear yourself think in the dining room. It did sound as though someone had turned on a monster electric fan on top of the metal roof.

When we come out from dinner we see that this, indeed and at last, is *the rain.* The dry wash outside has filled with water and rolling stones so that noise is everywhere, inside and out. The water is the color of milky tea and is roaring down, carrying pieces of mountain along. The wash is so full that we can't drive our cars out by the regular way, so six of us bundle into the station van and drive out the back gate.

We come out of the canyon and drive the long slow road down the bajada, the descent, the basin slope. There are toads in the road here like animated gravel.

We stop and get out to look. They are western spadefoots of the New Mexican variety: *Scaphiopus multiplicatus,* two inches long and the color and texture of a lump of potter's clay. They have reddish bumps all over like glassy imbedded beads. Their bulging eyes have vertically slit pupils; the mark of a spadefoot. Touched with a sneaker toe, they hunch and shut their eyes,

going more clay-lumpish than ever. I pick one up. It has spades
on the undersides of its back feet, hard black shovels with the
texture of tough fingernail. It scritches at my palm, doing its
best to scrape itself backward out of sight.

I put it down well off the road where its clay-lump camouflage works just fine. We drive on. The toad has left a smell on
my palm, an oily whiff of raw peanuts. With quick-digging
equipment, clay-lump camouflage and clay-lump behavior, and
noxious-tasting funny-smelling secretions, the spadefoots come
well defended into the desert night.

They need all this. They've come a long way down from the
upper bajada where they've been buried, a yard down under, for
eleven months or more. At the end of the rains they'll go up
there and bury themselves again and go dormant and cold, like
tubers. When the first hard rain brings them down they come
steadily and fast, for toads, but hardly fast enough to escape
coyotes, foxes, snakes. Without their armory they wouldn't
arrive.

We stop every quarter mile or so and turn off the engine to
listen. The sky is rolling and dark with quick clouds, with
thumps and flashes of storm. Now we hear a distant rattling, a
closer roar, bleatings, blurts: the mating songs of spadefoot
toads.

The air is thick with smells of spice and herb — the pungent
potpourri of creosote bush, the effluvia of desert rain. There is
an undersmell, too, of soaked silt, clayey, like a potter's shed.
Gravel and water strew the road. The van lurches off the tarmac
onto a dirt track gullied and pasty with wash and wet. The
ditches on either side are filled to their brims with silty water and
more pours down. Drowned shrubs and grasses and weeds poke
from the water, wag in the current.

The van slows; we pull up the hoods of our rain suits and grab
flashlights. Rumor has it that three desert spadefoots can be
expected hereabouts, their ranges overlapping just in this corner
of country: the clay-colored westerns, the darker and more

mottled plains with their snub noses and bulging foreheads; the larger Couch's with their green and yellow camouflage.

They should all be here, they should all be singing. We stop the van — the road ahead is a river we don't dare cross — and we open the doors and are assaulted by the noise.

The air vibrates with the westerns' rhythmic rattles as if someone were beating wooden drums impossibly fast. It's an oddly African sound, woodenly musical, almost a snore. This chorus is punctuated by lorn bleats, quackish blats at intervals as if a goat were begging; the big Couch's.

We shine our flashlights into the ditches. The water is opaque, pale, rilled by flow, scrimmed by bright streaks of rain. There: a big green and yellow Couch's upright in the weeds — and an-

Couch's Spadefoot Toad
Scaphiopus couchi

other and another. There: a western already paired, a two-frog unit, the smaller darker male riding his female. There are more, more; more smug silent pairs bobbing. Every Couch's is a different variation on the camouflage pattern: browns, roses, lemony yellows, olives, real greens. The males rock in the ditches, blowing their whole front ends up like balloons and rocking back again as they deflate with a loud *wooaaaaaeh*. Like those of all spadefoots (spadefeet?), their eyes seem much too big for the smallish toad they are.

The water that washes over the road carries toads with it. Mated pairs go downstream, rolling like big stones, pale bellies up and then flipped and up and flipped again until the whole is swept into the far ditch and through the drowned weeds and off into the dark. The males hang on all the way in mute desperation, or ecstasy. It's hard to tell.

A series of short grunty bleats — *weh weh weh weh weh* — announces a lone plains calling. We swing to find him. There he is, grasping weed stems in both front feet, his throat swelling.

We're waving our arms and wading up and down and shining our lights, but all our hoopla makes no difference to the toads. If we come too close to them they duck and scoot off and begin again two feet away. I am soaked to the knees. We all are.

Dave has flagged off a section of ditch and is writing busily in a notebook, shouting instructions to his hooded assistant, his hands juggling pen, flashlight, vials. The toad parasite that he's after is as faithful to rain as the toads are. Like their hosts, the parasite larvae have one night a year loose in the water, one night in which to accomplish their trajectory.

This is a one-night boogie, a once-a-year treat, like a trip to the opera or dinner in town. By dawn the chorus will fall silent.

Tomorrow the singers will be gone.

The spadefoots aren't the only ones here. Their chorus is backed by the calls of the great plains toad, whose Latin name, *Bufo cognatus,* means the toad-who-knows. He is classically toadlike, beefy and dark and liberally warted like a pocked volcanic stone. His throat sack is thin, membranous, pink as a bubble-gum bubble. His call is something else again. Upright in the drowned grass, he lets loose with a high puling whine that assaults the ears like a rock-band amplifier gone into overload. We grin at each other and block our ears.

The tiny green toads are singing, too. Their Latin name, *Bufo debilis,* means "weak toad," which I think is unfair, but they make up for any unfairness by being beautiful. They are a rough, bright, iridescent green with coal-black traceries, the male's throat bell a clean pure white. The overall effect is of a piece of animated jewelry. They are little more than an inch long and are master ventriloquists. We found one finally by a kind of triangulation; two of us waded together through the drowned grass, each sure all the way that the toad in question was in opposite directions until we found him between our legs, his throbbing trill filling all the space of air, rain, road; everywhere but where he was. Even when I saw him I heard him singing two yards off. How does he do that?

I spent the rest of the night scouting every road and track I could find in search of noise; puddles, toads. What I discovered was that almost every puddle I found had only one kind of toad in it. The westerns stuck mostly to uplands, the Couch's farther down, the plains at odd intervals between. That first place was a strangely ecumenical ditch.

Most of the toad puddles were cattle "tanks," scoops bull-dozed in the ground to catch rainwater for cows and steers. The toads won't use water that's been standing around awhile. Water that's been standing around is full of bullfrogs. Bullfrogs are a kind of amphibious rat; they follow human activity around at the expense of native amphibia. Their big tadpoles can gobble spadefoot tads as fast as they hatch. Standing water is full of water beetles and they eat tadpoles, too, so only fresh rain pud-dles are good for spadefoot business. In the old days they used buffalo wallows. What will they use, I wonder, when all of this is mini-ranchettes?

That there are toads in puddles isn't so fantastic. Toads have been designed to take advantage of temporary wet. What I wonder about is this: this whole webbed and splayed design has been made for puddles, but these spadefoots spend just a few hours a year in puddles and that's all. As we would use a voting booth, maybe. Or an appliance store.

Their tadpoles are quick, too; very quick. They can hatch and grow and metamorphose to new young terrestrial toads in eight to twelve *days*. In order to grow like this they eat a lot and they eat all the time; they'll eat algae, fairy shrimp, even each other. The cost for temporary pool use here is steep.

When the toads finish singing and mating and laying their eggs, they get out of the water and bury themselves. They greet the dawn rootlike, gone. Their burrows are scrabbly, like a skunk's grubbings in a lawn. When night comes or when it rains again, the toads come out to eat; they eat termites mostly and they eat a lot: up to half their body weight in one night's feast. Six feasts, more or less, is all they need for the year. With the autumn cool they migrate upslope for the yard-deep burial, the long sleep. They are awake and doing for a month a year, and only at night, even then.

If spadefoot song and sex and childhood all seem a little short, the truth is that they are not always short enough. I have seen cereal-bowl–sized puddles in muddy cattle tracks, all that was left of a tank that had been filled to overflow three nights before. Each bowl seethed with tadpoles. Three nights had passed since the air here vibrated to the male's snoring drumroll, when dozens of paired eye-lights shone back at me as I shone my light at them. Now you could scoop the tadpoles out of those cattle tracks like spaghetti out of a cookpot. One more rain would make it all into a pool again, but even with song and dance, the coloring of costumes, and all the prayers in the world, the rain does not always come.

Burrowers

I was on my way home from the desert last night when a mountain lion jumped from the brush beside the road and ran on in front of me for a long way. His tail was lifted as he ran, like a kitten's. I ran on after him until I was breathless, and I ran for no other reason than that he was running. He loped along with an easy ears-back canter like a cat when it crosses the porch with its tail in the air.

When he went out of sight, I stopped, panting, cursing my pack with its bobbing weight. The cat seemed then to have been more omen than real: a lion, after all. Power peeking from the dark wings of the world.

A cloud shut off the blue-black of starred sky, misting the eastern mountains with rain. Half a mile after the lion I stopped again because the canyon echoed with screeching brays. A pair of skunks were caterwauling on a slope below the junipers, mating in a contortion of black and white hair and howl. They were oblivious to me, but I backed away from them slowly as though I had been indiscreet.

When I got to the station, there were more skunks around the trash bin. They had an odd lopped canter as they went off — their usual pace is a sedate pad-pad — and they looked abashed trotting away with their noses to the earth like hounds.

Two great horneds were hooing in the high country. I had to shoo peccaries out of my path, stamping and saying "HAH," but even then they snortled at me and hardly budged, and I went through them snortling back as if their language made all the sense in the world. As I came in reach of my own door, a mule deer bounded past and I jumped, said something unprintable, and tripped, and my pack and I went down sprawling on my cabin steps.

I felt like a fool. Perhaps we are all fools here together. All the clocks are wrong.

It was almost dawn and I was hungry. I seem to be hungry all the time. I ate tortillas and cheese and drank a beer — durable food, pale and gravelly as the world. I shed my snake-guards and boots and jeans and sat bare-legged on my cabin porch, and watched dawn color the sky; then I went in and pulled the curtains and worked on my notes until noon.

In the afternoon I sleep. I have yanked my mainspring and turned it upside down and stuck it back, and backward it coils silently and lets go.

I pull the desert in after me everywhere, a presence that won't let me go anymore. When I'm there at night with my headlamp on, I am ignored. This much I've discovered now about night work: my own invisibility. It isn't only the deer, the peccaries, the kangaroo rats, the lion, the skunks; down on the desert flats the antelope jacks run over my feet or stop beside me as if I were a mesquite bole, their front legs fine and grooved as a deer's, their faces bony and aquiline as a Nilotic woman's. Come daylight, things change. At dawn the jacks dodge off with a scut and flip of black-tipped tail. At dawn the Gambel's quail bounce out of the washes like spilled marbles. Only in the dark can I move around out there as if I were in a house, their house or mine.

Last night I walked through the kangaroo rats' burrows as if I were a minor god, a striding presence to be ignored as all gods are; my purposes inscrutable. I was there to watch, and I saw more than I'd expected to see. More than I had any right to see.

All morning I have tried to write it down and it's been hopeless; words run off the page like water on metal and penetrate nothing. Some old memory has been jogged and interferes; some image wants out here and is as insistent as a strain of music: my notes blur in front of my eyes, papers are strewn about; none of the words convey what was. It is raining again and the lunch-bell is about to go, and I have to find my shoes: how can I begin?

As I go on here night after night, I find that something usually strong in me is coming unglued. Perhaps it's because my nightly encounters here are more animal than man. My separateness has become illusory. Perhaps that's it; I'm like the prisoner in solitary who adopts a mouse or a spider, or the hermit who discusses the Psalms with his goat. Maybe the ungluedness comes from living time upside-down, but I've done that before, and it wasn't like this. I find now that I have to hold on to myself, that sometimes I have to think to breathe. I know that I used to go from day to night and through each one as though I could eat up time without an appetite; I remember being pushed through light and dark as if the planet's travel were a simple thing, like breath should be. Now I feel . . . *hurled*.

In the afternoon the thunder wakes me up. I'm bleared with sleep, but no one can sleep through this. The thunder bams again and I feel beaten like the skin of a drum; every molecule jigs. OK. I'm awake. I'm up.

I'm out the door. I watch the darkness coming from the desert. I wouldn't miss it for the world.

It sounds like stones rolling. Already the first drops have patted down; the crumbled edge of cloud has rolled upcanyon and the cool front of air collides with stone, with other air.

There is an uneasiness in the trees. Then the rain hisses steadily in and the curtains billow behind me. The air smells of rain, of stones, as explosive as dust. Trees writhe. Eaves pour. There is a *crack* overhead and then a descending roll as if the mountain were coming down.

The storm brings dusk, a settled shadow. The curtains have calmed and now the rain falls steadily. The thunder makes troubled faraway mutterings, the trees have gone quiet; already the little leaves hang shyly at the ends of their branches, silvery drops held on their tips a moment before falling. The leaves are green. People are running across the grass and the grass is green too. This is not the first rain and it will not be the last.

The rain is the benison for which the Hopi people once danced down their mesas with snakes in their mouths. The snakes had been feted and fed with cornmeal and were released in the cardinal directions to carry the Hopis' rain prayers to the spirit of the underworld. So: the prayers have been carried. Last night there were two Mojave rattlers among the kangaroo rats' burrows, and they had to be removed on the end of a stick.

Down there in the flats the rain falls — not into dust and gravel and bare stone anymore, but into the seed-leaves and roots of sprouted grass and wild flowers.

Down there the desert has filled and greened faster than any spring. Between the rats' burrows there are rainwashed rills of debris — sticks, stems, leaves — and two days after the first rain these rills were filled with sprouting plants.

Last week the rats cleaned their storerooms of chaff and mucked out their latrines, and the rain washed the trash into the rills and the new growth.

In weeks there will be a waving surf of green where there was dust before. The stones will show in the mobile green like stones underwater. In the autumn drought the leaves and flowers will burnish and the rats will gather seed. Seed will be packed by the cheekfuls into the freshly cleaned storerooms.

The little gray Merriam kangaroo rats are nine or ten inches long and all but three or four inches is furry tail. Their tails are bristle-tipped like paintbrushes. The name "rats" seems wrong here, conjuring urban visions of cat-sized monsters with beady eyes; kangaroo rats are nothing like this. They are engaging animals, more squirrel-like than anything, with wide domed heads and great black eyes and round ears tucked back — like hamsters, maybe, or chipmunks, except that they're larger than either of these. They have a quick hopping-popping motion that is as jittery as a wind-up toy when they're on their way fast, a kind of kangaroo bounce at rackety speed. When they're foraging, or sniffing the ground, or gathering seed into their cheek pouches, which bulge like sacks, they lippety-lip like rabbits.

The Merriams are the smallest kangaroo rats here. They claim home territories that they've pocked with burrows and seed caches. The territories are lambent, flexible things; the males overlap with females and with one another, though the females don't overlap too much with others of their own sex, as a rule. The females mate with neighbor males, with those they know.

The Merriams share this valley with other kangaroo rats. To the east is a colony of Ord's, which are larger but look very much the same, and just to the west is a colony of the big bannertails, which are fifteen inches long, weigh three times as much as the Merriams, and have white tips on their tails.

In daytime they all plug the doors of their burrows with dirt and go in to keep cool. A beetle, a lizard; that's all that moves here in the light.

Last night there was no moon and no rain either and the Merriams' burrows were plain enough, each under its bush of mesquite or ephedra or creosote. The doors were open, and the occupants were out.

From dusk to dawn rats pop and scoot from bush to burrow to bush and it's busy enough in there that after twenty minutes one isn't jumpy anymore oneself, and it's like being in a village, and all this leaping and dashing is curious. One wants to know how it works; this is what Jan wants to know, too.

Jan Randall began her study of kangaroo rats' behavior eight years ago. For the first two summers she worked by herself and at her own expense. No one had studied the rats' behavior before, except in a lab. With the exception of squirrels and ground squirrels all rodents are nocturnal, and no studies had been made on their natural behavior in the wild. No one believed that it could be done, or thought that Jan could do it, or paid much attention one way or another.

Jan wore snake-guards and learned to pick Mojave rattlers up on the end of a stick and move them out of her territory. Her published papers, in the understated way of such things, give clues to the amount of time she spent: "I observed two matings in each species during over 1000 hours of behavioral observations. . . ." ". . . in approximately 250 hours of observation of mounds at night during June and July 1985 about 25 aerial passes by owls were observed . . ." and so on. Her data were meticulously compiled, her conclusions fresh. After two years she began to get plentiful funding, assistants, invitations to give papers at seminars and meetings. She has been here every summer since. She has been here all of this past year on a grant from the National Science Foundation and the National Geographic Society. She has been studying the big bannertails, which have their mounds in the open gravelly flats, and the little Merriams who build their burrows here, in brushier country.

If animals live in more complex societies than bannertails or Merriams do, then they've had to evolve the systems to do it with; and all these systems rely on communication. There have to be totems and taboos, rewards and punishments, machineries of social rank, loyalties to family and clan, and armed and unified defense. We do this. So do chickens, gorillas, hyenas, ants.

The rats are at a basic stage in the social business. Each of the rats is sovereign of its own mound or burrow, and this is all they need for defense against coyotes or owls or midday heat. The brief foraging territory and the underground labyrinth of nests, latrines, and storerooms is all they need for their comfort and sustenance.

Aside from the odd night of mating and a few weeks now and again for rearing young, each burrower doesn't have much to do with the other rats around the place. This is hardly a society. Or is it?

That is what Jan is finding out here. She is quantifying the rats' communication and social behavior; the primal machineries of getting along.

The rats see just fine in the half-light of desert night, and moonlight doesn't bother them, though wind does; wind confuses the sounds and smells that make up the greatest part of their sensual language. When it's windy on the flats, the rats stay home.

Otherwise they're out and about all night. They forage, eat, dig. The bannertails bathe in the sand at the edges of their mounds, leaving their scents in the dust. They drum their mounds with their feet; a long-range claim, like a bird call. If a snake comes too close, they drum at it, telling the snake: I see you, no chance you'll catch me now. The snake understands this low-frequency language. After a while it slides away to find less wary game.

Any foreign rat on a home territory is cause for alarm. If an intruder comes too close, the owner pops in the air and takes off in pursuit, for two or three yards or more, before going home and back to business. If the intruder is persistent and challenges again, he's chased again, and once in a while there are rolly scufflings before the foreigner gives up and goes away.

The bannertails and Merriams have their differences — in building style, in social style, too — but there is one thing they have in common. Only at mating time are intruders allowed. At mating time there are a lot of "interactions," as they say in the behavioral trade, and this makes for good data. This is more than the mechanics of claim and raid and trespass: this is sex and war.

The Merriams mate in both winter and summer rains so that their young will be born with both seed harvests; the rains are the cue. For some nights now the male rats have been gathering on the females' territories, smelling at urine marks for signs of estrus. They have all been firmly chased off. Until now. Until last night.

Last night Jan was away. Her assistants, Mark and Tammy, were alone out there with their tape recorders and their lights.

Tammy was burdened with a radio antenna and the black box to run it with. The radio-tracking work had started a few weeks before and it was useful, but the gear was a nuisance.

Some of the rats had radio transmitters implanted under their skin. The transmitters made awkward-looking bulges but didn't seem to interfere with business. Each rat had its own transmitting frequency. With antenna in hand and dial tuned Tammy could find out where each rat was, at any time, underground or above.

The burrows had been checked in the daytime to find out who was sleeping where (a certain amount of burrow-switching does go on), and it was easy at night to find out if so-and-so was home, or on the lam, and if so where they'd lammed to. With the radio-tracking gear, and with the color-coded eartags that all the rats wore, and the knowledge gained from weeks of observing each rat's appearance and home territory, Tammy and Mark made order and sense out of what seemed to me at first to be a jumpy chaos.

"There's a mating in progress here," Mark said when I arrived. "Stay away, please."

I stayed put but still. Mark was by a burrow four yards off. Rats popped and circled at his feet, and he spoke quickly, low, into a tape recorder:

". . . male 279 approaches her, nose-anal nose-anal circle circle, nose-to-nose . . . she moves away, he's driving her, driving her, she runs away . . ."

A rat zips over my feet, sits up, scoots.

". . . ragged-ear male approaches to the west; male 279 sees him, nose-to-nose, circle circle . . . female 234 foraging three yards to the south . . ."

After half an hour I know who's who: lop-eared female 234 with one gold eartag and a transmitter bulge on her rump; larger darker male 279 with his single red eartag and lopsided transmitter; pale, small, transmitterless, and tagless ragged-ear male:

". . . ragged-ear male approaches around her mound to the west, 279 chases him, circle circle jump, 279 moves off, ragged-ear male by burrow . . ."

I orient myself to the territory: the clump of mesquite and ephedra with the burrow underneath it, with its two nest-holes and dents of seed caches; the smaller hole to the west under another mesquite but part of 234's dominions; the dry wash, narrow as a mountain rill, to the east.

Rats whiz, circle, zip off, hop back, forage, chase. We lose sight of one or the other, then of all of them. Mark calls Tammy over and she brings her antenna. The pingings are chirpy, strong, right there at the hole under the mesquite bush. Female 234 is in there. So is male 279. Almost certainly they are mating now.

". . . ragged-ear male approaches the hole and sniffs and goes away. . . . It is now ten-twenty-six . . . male 279 and female 234 have been in there for twelve minutes . . ."

One rat emerges and scoots off, then another. Which and where? There are circlings, chuffings; verbal annoyed natters like softly tearing fabric. Rats zip and disappear, reappear.

Ragged-ear loses most of the circling hopping battles with big 279, but he comes back, again and again. Suddenly he catches up with 234 just to the east of the mound and there is circling and nose-to-nose and nose-anal and ragged-ear drives her and mounts her, clasps her, lifts her up, thrusts hard and thrusts harder, both rats hopping together . . .

". . . he's still thrusting. Still thrusting. Thrusting harder now . . ."

After five minutes 234 gets loose and runs into her burrow and plugs the door with dirt.

Ragged-ear hops up and scratches at the door. He goes away. 279 comes and sits by the burrow and eats a piece of greenery and grooms his feet.

Half an hour passes.

Mark goes away to work with Tammy. Another mating is in progress forty feet away. Their lights flash like stars, there is a low mutter of behavior language, and I'm alone. I watch the closed door, the empty burrow, the shadows of the ephedra bush; but I do not watch them well enough.

Suddenly chaos reigns. I've missed her exit but 234 is out, her hole is open, she's gone. Rats are everywhere; there are circling

Merriam's Kangaroo Rat
Dipodomys merriami

chuffing battles and chases from bush to bush to wash to mound and round and back again; at last I see 234. There she is, lop-ear and gold tag, prone under a bush by the wash, and a strange new male is mounting her . . .

"Mark!" I call. "Here. Quick!"

He runs.

He pants into the recorder: ". . . female 234 in lordosis, three meters east of her mound. . . . It's eleven-o-four . . . male 507 is mounting her, he's not thrusting . . . no penetration. He's grooming his feet . . ."

Where are the other rats? 507 and 234 are alone in the wash. 234 lies on her belly with her front paws out and her nose in the air. Her paws are white and her eyes and whiskers are black and she looks very appealing. 507 climbs up her back, he's over her left leg, he squeezes her and chews at her neck and then . . . she wiggles. He falls off. He climbs up again, over her right leg . . . she wiggles again. He sits up and grooms his genitals.

My fists are clenched as though I were rooting for the home team. Mark coughs. His fists are clenched too.

"I think he's probably ejaculated already and can't do it again, so soon, I mean," he whispers.

"Maybe he's just young," I whisper, "and doesn't know, you know, the ropes."

"Come on, 507 . . ." Mark hisses.

We watch.

"Aren't we just being anthropomorphic, Mark? I mean, they're rats . . ."

He looks at me as though I were a crazy woman.

"What's wrong with that?" he whispers.

Psalms and the goat, I think to myself, *the prisoner and the mouse . . .*

We watch.

234 lies prone with her white paws out. She lifts her nose. 507 grooms his genitals.

After a quarter of an hour of this 234 gets up and runs back to her burrow. For a few minutes nothing happens. Then 279 appears from somewhere and puts his head in her open door but she shoves him out, natters at him, and turns and plugs the door with dirt; and that's the end.

It's midnight; this has been going on for four hours. My eyes have been on the ground. The mound and the wash and the ephedra bush and the mesquites have been the world. They are the world. A sift of cloud has crept overhead — a big winged cloud shutting off the blue-black of starred sky, misting the eastern mountains with rain.

Valley of Snakes

He says he's glad to have the company, so one night after dinner I go out with Joe Mendelson to look for snakes. It's pouring again and we get into his truck in a jumble of streaming rain gear, notebooks, flashlights, packs. At my feet is a pile of pillowcases and cloth bags. On the seat between us are a notebook, a pencil, and a clock.

We roar into gear as if we're already late. The headlights sweep the camp so that cabins and people spatter past like black-and-white stills, the wipers sweep curves of streaming windshield; we splash and bump over the wash and we're off downhill. We're going to the San Simon and San Bernardino valley road, the ribbon of blacktop that runs to the border of Mexico.

In the back of the truck are a modified golf putter and a broomstick with a hook on the end. There's a cooler and some plastic bags for dead snakes and a "hotbox" for live ones. The hotbox is a five-gallon plastic bin with a hole in its lid. A square of steel mesh is glued over the hole. A warning is written in black letters all around the hole: "DANGER LIVE VENOM-OUS SNAKES DO NOT TOUCH."

Joe was once a student at the University of California, but even then studentness was secondary to his work with snakes. He's slim, blond, tall, and has an intensity about him so that he looks as if he were always leaning into a wind. He's a volunteer at the station, and will be here for six weeks, but his volunteerness is secondary, too. He has to spend four hours a day painting buildings or working in the kitchen to earn his bed and board, and the rest of the time he can do things with snakes. He has been doing things with snakes all his life, and there isn't much about them that he doesn't either know or want to know. This part of Arizona is famous for snakes and he's been here for less than a week. The truck crawls down the canyon road, the rain thins, the windshield wipers saw against the spattered wet, and he's on the edge of his seat.

Suddenly there is a pale line in the road like a section of gray hose. We stop. The rain has stopped and so has the snake; we get out of the truck and then it moves. It slides across the tarmac in a straight line, heavy-bodied, its color soft like a line of symmetrical reddish stones in buffy sand; the scales pointed, keeled, dull. The black-and-white-striped "coon" tail with its pale rattle is cocked above the road like a warning finger. The snake's tongue flicks; he is trying to get our number.

That tongue is a collector of airborne molecules. He's collecting us: sweat, breath, secretions. In the roof of his mouth are parallel grooves and pits and the tips of his tongue flick and fit in there. These pits are the Jacobsen's organ, the vomeronasal organ, the VNO for short. There the molecules are collected, chemistry converts to nerve signals, and these are shunted in.

The VNO is central to a snake's knowing of the world. Without it he won't even mate; it's not the sight but the VNO-sense of a willing female that means arousal to him. It connects with the deepest brain centers. It is telling him who we are.

We look at him and he flicks at us and I think about this: 300 million years ago, more or less, we and this snake had a common ancestor. It was a four-footed animal with a VNO. The pits in those fossil skulls tell the story; here is this brain-socket to plug in the world, old technology developed upward from a fish's nose. Time squirrels things away. Even now all four-footed vertebrate embryos have VNOs. Primates, birds,

Western Diamondback Rattlesnake
Crotalus atrox

whales, bats, all have them in utero or in egg, then the twin pits
are buried in development and are gone; vestigial quirks in the
unfolding of form. I had them once.

Now I have the lights of the car.

"Western diamondback," says Joe.

He looks a lot like a Mojave, but the black bands on his black-
and-white-striped tail are wider than a Mojave's, and his coloring is dustier. He does have a reputation for nastiness. A
diamondback has potent venom and a short fuse. He needs this;
his life here isn't safe. Skunks, foxes, coyotes, wild cats, owls,
all eat snakes whenever they can. A horse or a bison hoof can
smash his skull, crush fishbone-thin ribs. He confronts these
dangers with his own tools. Like the Lone Ranger and Tonto
with their ears to the ground he can hear the thrummings of
approach, and he bunches and sounds his warning: Here I am,
the dusty one with the foul temper, and I can hurt you so go
away.

This one is coiled now and doing his rattling, putting on his
show. He's in the road, though, and cars are coming. We wave
two cars down and guide them past, then Joe takes the hooked
stick and pushes the snake off the tarmac and into the safety of
the ditch. He's a buzzing clench the size of a hubcap. He's holding things up.

Back in the truck cab Joe writes it down: "*C. atrox* A.O.R.
1.2 mi. E Portal Az. 7:30 P.M."

Alive On Road. First entry. Check the mileage now: on
we go.

The next one's a garter snake, limp on the shoulder of the road,
dead but hardly damaged.

"*Thamnophis cyrtopsis*," says the entry, "D.O.R." Dead On
Road. Time, road, mileage: data.

She's slim, dark, a lowland hunter of tadpoles, frogs. One
night I saw one like her swimming in a rain-filled ditch, head up,
writhing quickly. The wonder is that they don't just sink, that
snakes swim so well. She's darkly checkered and has a yellow

blaze the length of her back; the blaze widens to orange at her neck. She's delicate, cool, limp. She goes in a plastic bag with a tag, and into the cooler, and that's all.

Another D.O.R.; brakes squeal, doors slam open, we're out. This one's a gopher snake, lying crooked in midroad like a stomped piece of wire. He's been run over twice at least and is beat up but he'll make an OK specimen so he goes in a bag in the cooler, too. The gopher snake and the garter snake and any other D.O.R.s we find will be packed in preservative and shipped to the Vertebrate Museum in Santa Barbara, where Joe works full-time now as the manager of the herpetology collection. He's worked there since he finished studying, though working there is studying, too; just as much as collecting is. Next year he'll be off to graduate school at the University of Texas. For Joe it's all part of the same.

He tells me about gopher snakes. The gopher snake is one of those ironies. One of those maddening things. A gopher snake is a harmless eater of rodents and defends itself by imitating rattlers. It has the same coloring and the same diamond-shaped mottlings as a Mojave or a western. It coils when it's threatened, it vibrates its tail and makes a hiss in imitation of a rattler's rattle. It does fake strike-lunges with its head. This act fools just about everyone. Unless they're very wise or very foolish all snake predators steer clear. The problem is that gopher snakes fool people, too.

People don't run away from rattlesnakes. They kill them.

Joe is on tender ground here. He leans forward over the wheel, we grind slowly over the road. The gopher snake can't change fast enough to shed the colors or habits which millennia of selective process have planted in its genes. Most people can't change fast enough to learn the difference, or to care about learning the difference.

This is where Joe works, on the fringe of ignorance — not only his own but other people's. When he's not studying he's teaching. People call the Vertebrate Museum when they have a snake on their lawn and then Joe goes out to take it away — to protect the people from the snake and the snake from the people, and to teach while he has an audience. Bring out the kids, neighbors, Pop and Mom; let them see the beauty here, the colors, let them see that this snake can be handled with respect, that this snake is harmless, and that this is how to tell the difference. Let them see that a kingsnake will twine around your grandson's arm as cool and smooth as a black-gold chain, and lick the crook of his elbow, and make him laugh.

Sometimes when Joe gets to the place, the snake is dead, its head chopped off, its tail chopped off, the kids all standing around saying Yuk, saying Eeeeew. A lot of the time the victim is a gopher snake. It's no good his saying anything then. No one is ashamed. Someone's a hero. Mister, there's little kids around the place, can't afford to take the chance. It was a convincing fake. A deadly defense.

Joe tells me that gopher snakes are nocturnal now, in summertime, but in spring and fall they may be daytime hunters as well. They hunt just fine day or night. What's important is temperature, humidity, sexuality, degree of hunger, availability of game, and other unknowns and unknowables, so that in answering questions or even asking them one is struck full by the stretchy complex web of what being here must be, for them.

Rattlers switch, too. They are more nocturnal in summer than in spring or fall, desert rattlers are more nocturnal than rattlers elsewhere, lowland types are more nocturnal than upland types, adults tend to be more nocturnal than juveniles; and so on and so on. Here we are then. Summer, night, lowland. This is rattler time. Here. Now.

There are other snakes in this desert that hunt strictly at night, or live and hunt deep in piles of rock or vegetation from which they emerge only in the safety of the dark. Joe reels them off: longnose snakes, glossy snakes, nightsnakes, black-headed snakes, leafnoses, shovelnoses, blindsnakes.

The blindsnakes are what he particularly wants to find, and it won't be easy. I saw one once on the desert flats after the second rain, and I thought at first that what I was looking at was an earthworm — same width, same color — then something clued me in. The patterned gloss, the S-curve motion, the fact that it didn't fatten and thin as an earthworm does when it's unching forward. No, this was a snake; tiny, fast, with black spots like pinpricks where its eyes should be.

Blindsnakes make a living by going down ant holes and lapping up larvae and eggs. They don't need eyes down there, they find their food with their forked tongues in the dark, just as they find each other when they need to. One of the ant researchers told me about seeing a blindsnake go down a harvester ants' hole one night. The daytime harvester ants, the *Pogonomyrmex* clan, are big and red and they bite hard. That blindsnake whooshed out of the hole with a dozen Pogos latched onto him like skiers on a rope-tow. He wouldn't make that mistake again.

The blindsnakes are most active now, in summer, during and after the rains. Now they're out and hunting. With the size and naked color of a piece of cord, they'll be difficult to see on the road in the dark. This doesn't bother Joe.

We have our low beams on, we're hunting, too. Joe anyway has a "search-image" — a mental picture built up from long experience in road hunts — and he calls out "*Thamnophis!*" or "*C. scutulatus!*" as we jam on the brakes, having passed the D.O.R. twenty yards back and at thirty miles an hour in the rain.

We make mistakes, though, especially since some of what we're looking for is very small. I make more mistakes than he does but he stops anyway for every "there!" I yell. We stop for a bottle, a paint blotch, wire, sticks, string, one barn-owl claw; but most of what we stop for is snakes. We do stop for spadefoot toads, too, and great plains toads and a smattering of frogs. We write down the time we find them, the mileage on the odometer, the species, the condition of the kill — if it is a kill, which it is, mostly. We are careful to hold the dead rattlers by the back of the head. Their venom is still lethal, their hypodermic teeth are still sharp. All it takes is a scratch. A lot of the snake D.O.R.s look perfect, hardly touched. There is a patch of body gone soft, a crooked jaw, but that's enough. Delicacy again; fragility cloaked in cool inscrutable scrawl.

Altogether we find two sonoran gopher snakes, four blackneck garter snakes, three checkered garter snakes, one hognose, two Mojaves, one four-foot western diamondback (plus that first one) and a massasauga, and two kingsnakes. We don't find a single blindsnake but there will be other nights, and this one is busy enough.

It's strange. My snake-shakes are gone. I leap from the car like he does, sometimes faster than he does, and I stand foolishly close to lethal rattlers more than once. I've found the antidote at last and it's simple enough: companionship.

We collect only one snake alive — the massasauga. Joe is excited about the massasauga. It will be refrigerated and shipped to the museum and kept alive there on display. Collecting it isn't easy, though; the snake isn't big but it can be deadly and it's quick — it's S-writhe whippy as a shaken rope, a plucked string — and it's pouring rain again and the cloth bag is there on the road and the rain pounds it flat and the snake S's in and out again quick as light. The temptation to reach down and hold-the-bag-open pick-it-up or use-your-feet-to-push is terrible but at last it is stick-juggled and jimmied in there. Joe drops one stick over the bag's mouth and rolls it up rolling-pin style until the snake is down at the closed end, then picks the bag-mouth up still with a stick (hands nowhere near) and ties it shut, and still with the stick picks the bagged snake up and sets it into the hotbox, gently, as one would set a new-baked pie in a basket.

We are soaked to the skin, we turn on the heat. We go on. Cars pass us, trucks. The road is line-straight on the valley floor and they come along fast. More than once when we are out on the midline looking at a snake we are warned just in time by the hum and rising lights of a pickup, and we run. When the big trucks roar past with constellations of cab lights blazing,

"Specimen-maker," Joe says, and we laugh. It isn't funny, really. We are on that road for five hours, though, and things get funny late at night, things that aren't, really, at all.

Between snakes, Joe tells me about skinks and sand skinks and Borneo earless monitors, all near-legless lizards who hold clues to how snakes became what they are. Legs are in the way when it's small places you want to get into. Small places are good to hide in or hunt in. Once you are legless, the classic reptilian side-to-side wriggle becomes S-curve writhe. Eyes are not much good underground; the ancestral snake's eyes were reduced to nearly nothing. The VNO was sense enough down there.

Somehow, down there, snakes survived the mass extinctions of the Tertiary. I see them like a pinpoint of light, persisting in whatever time that was, a Terran dark age; afterward they came up and radiated into other niches — I see that pinpoint spreading like a star — and their eyes re-evolved but could never develop what had been lost. Snake eyes have no fovea, no screen on which to focus an image. Snakes can see movement, not form.

It was just about then that we found the four-foot western diamondback, as big around as my arm, twisted on the road like a crushed pipe.

That's one I won't forget; not soon; not ever. There was a whole cottontail rabbit inside that snake. The rabbit was lying nose first and legs outstretched as if it were being born instead of digested when the Chevy struck.

A diamondback hunts by ambush. He smells out rodent and rabbit trails and hides beside them, waiting for the telltale vibrations of hoppity-hop. He has no visual "search-pattern" but he can see whatever moves; the lippety-lip, the ears' wag-wag against the lighter sky.

His VNO tells him what kind of thing this is, this cottontail coming toward him in the dark. He homes in using the twin pits of his infrared sensors, he "sees" the cottontail in stereoscopic infrared image almost as we would see an image in light. That heat-shape, wavy and fluxing as a held flame, comes closer; then close enough. The snake erects his fangs with a special bone lever rooted in his jaw. He strikes and pumps venom and pulls back in a quarter-second punch. The cottontail rocks in his tracks, hops, wanders, stumbles, falls over, kicks once, dies.

The snake waits, licking the air. Then he follows the trail. Not the trail coming, the trail going off. He finds the cottontail, flicks his tongue over it, knows: this is mine, this is good. Starting at the rabbit's nose, the snake begins his swallow. This takes a long time. It is like something being born; convulsive and slow.

The other one I won't forget was the last. We were almost home, and I saw nothing at all, but Joe did; and he braked and jumped and was back again with it in his hand in less than a minute. He poured it into mine. It was a newborn garter snake with a head the size of a fingernail, a tail trailing into a black thread; it was limp as a thread, cool as metal.

It had a high gloss, like enamel. The orange at its neck was bright as fire. A tire had just ticked it; the lower jaw was crushed, that was all. It was newly unwound from the egg and newly gone.

We had begun to wonder on the way home. We'd been over this road on our way down but there were a lot of D.O.R.s on the way back that hadn't been there before. It had rained again at midnight and maybe that had kicked a lot of snakes into motion, like someone letting out the clutch.

In India the kraits and cobras come out after the monsoon breaks. In the Sudan the vipers appear, suddenly, on doorsteps, in the rare rains. Rain brings the snakes out here, too.

The problem with the rain theory was that there hadn't been many cars. Perhaps a dozen had passed us altogether. It was well past midnight and this was ranch country where people bed down with the cows, mostly. But there were so many snakes dead on the road! We couldn't understand it.

If a dozen trucks and cars had killed all these, then how many had crossed the road and hadn't been killed? What kind of percentage were we dealing with? And — the thought came to both of us at once like a bomb going off — we were talking only about *the road*. What about the desert, then? How many? How much?

Joe banged the heel of his hand on the steering wheel.

"Oh my *God!*" he said.

We were quiet the rest of the way.

Up in the canyon we saw no snakes at all. In the autumn some of the desert snakes would come to the foothills to den but now they were down there, behind us in the dark and as far as we couldn't see; as far as we would never know. Long handfuls of cool flexing color, friezes fit for the mosaic of a Roman bath, all sliding, out there, in the dark.

Alien Ground

I've looked and looked but I haven't found a one. "A one" isn't what I mean, exactly. Ants come by the thousand or not at all, and a one is a fleck in the void, a single neuron unhitched and nowhere, an accident. I want an armored division, a bunch. I want army ants: a raiding column, an attack, a booty-laden homeward journey, sun spiders crouching at the column edges picking off stragglers; the whole kaboodle. Every time I see Howard I ask him:

"Where do I look for the army ants?"

"Everywhere!" Howard says.

"I'm looking almost everywhere, but . . ."

"Have you got good boots?" he said. "They bite."

This is not news. Plenty of things out here bite and I have faith in my boots, so I spend my nights walking on the desert flats. I walk wherever I can find a road that I'm sure I can follow home again. In this way I am like them; they follow roads.

They follow roads of their own making, chemical trails as clear to them as the ditch and fenceline and double-trail truck track is to me.

It isn't strange that the army ants are busy in the dark. Darkness is no problem. All ants find their way just fine in the dark of their nests, but the army ants' nests change around all the time; they live in nests stolen from other colonies or in bivouacs quickly cobbled up from worker ants twined together. The live tents are handy, and are quickly uncobbled again for next night's march. I'd like to see one of those: the live sizzle of anty architecture with all the workings hidden in the glittering bulge; the queen, the brood, the warriors. They must have ways of "finding their way around" that are utterly beyond me.

It's a successful way, whatever it is. There are army ants all over the tropical world. They're members of the subfamily Dorylinae (the word comes from the Greek word for spear); they all sting, they're all carnivorous, and they all stage heavy raids. They all have large colonies. The African Dorylus can have more than 20 million individuals; they are an awesome business, consuming more meat per acre per year than the local lions, leopards, and cheetahs put together. The *Neivamyrmex* genus, which lives around here, is more modest. These are the northernmost of the Dorylines, the colonies aren't so huge, and the ants aren't, either — though this is the other thing: most army ants come in variable sizes. One colony can have everything from specklike minims to big soldiers and everything in between. This is one of the ways that you can tell what they are.

I've got the message: army ants are here, and not just here. The *Neivamyrmex* genus is spread all across the southern United States, down through Central America, scattered through the Amazon basin, through the pampas of Argentina: everywhere.

They're nomads, like all Dorylines, and sometimes they're more nomadic than other times. When their brood is pupating, they stay relatively put, in bivouac or stolen nest, and new eggs are laid, and family business comes first. When the pupae hatch, the ants are on the march again, carrying the larval brood with them from camp to camp, raiding and feeding heavily as they go.

They raid other ants' colonies, and they feed on the larvae and pupae that they find there. They send scouts on ahead to look for likely victims. Scout trails branch and fan out from the nest across the desert floor. One of their favorite hits is the Novos, the nocturnal harvester ants that I'd watched at Silver Spring and that really were "everywhere" in the tracks and by washes. They're handsome ants, with slim waists and neat, glossy abdomens.

The *Pheidole* are another prime army ant victim. I saw them all over the place too. They are tiny and carroty red and they have a number of alternative nests, moving the whole colony from one to another at intervals and in orderly brood-lugging columns. This is what Howard calls the "MX missile ploy" and is their way of surviving army ant attack: routed from one nest, they have others to go to. Their columns on the move are flanked by the soldiers who have big red heads and who keep those heads up and jaws spread, taking their jobs seriously.

These are the two to watch, Howard says. I mark their colonies with sticks and check the sticks coming and going and coming and going again, but there are no signs of war. I begin to wonder if I understood correctly. Where exactly is "everywhere" if it isn't here? I'm beginning to feel foolish with my little sticks.

Howard Topoff knows army ants better than anyone. He is a professor of psychology at Hunter College in New York and he has studied army ants here for twenty years. He doesn't study them anymore because he's developed an allergic reaction to their stings. He's studying slave-raiding ants now. He's got four raider colonies staked out up behind the station and almost every day there's a raid up there at five-thirty in the afternoon, columns marching toward target nests and marching back with pupae from the raided colony. Next morning there's a heap of white shells in the slave-raiders' midden; the remains of a victory feast. The rest of the captured pupae are reared as slaves that do all the work of the colony, building the masters' nests, feeding the masters' young, cleaning up. The raider ants themselves do nothing but raid. The Centurion reigns, there is no Spartacus in sight, and the question is, still, with slave-raiders as with honey-pot ants, fungus-growers, harvesters, army ants: how is it done?

Once you ask that question, you can't go back. Ask the mayor of Chicago, the admiral of the Pacific Fleet, the chairman of the board: is organization simple? Ever?

Or Volga Vseslavich, the famous Russian enchanter. No one can ask him anything at all now, but sometimes when I'm alone on the flats in the dark, I wish I could. It was a long time ago that he had his organization problems, but he had talents we seem to have lost; powers of change that have drifted off, gone, like smoke. In his time, Volga Vseslavich was worried about the Tatars, the Golden Horde perched up there in the Caucasus ready to swoop down over Russia. The castle of the Tatar Khan was well defended, and the Khan's sentries saw nothing strange in the column of ants crawling single file through the bolted gates; but once they were all inside Volga Vseslavich turned the ants back into Russian warriors. At dawn, not a Tatar in there was left alive.

There are strange things out here, too. There are marvels here and now worth watching for. The ants' appearance of chinkless fit, of utter individual obedience to the greater good, may be enviable or awful depending on who you are; but there are, even in ants, things that don't seem to fit. There is room for the odd.

Howard has seen a lot in his years of nights studying army ants. There are things that he's seen in the course of looking for something else. Once he watched a colony of Novos collecting feathers. They couldn't eat feathers, but the feathers were stuffed and jimmied down into their nest. He wondered about that.

Days later the army ants attacked. The Novos boiled out of all their doors at once. Most of them grabbed what brood they could and ran for the hills — the refugee ploy: chaotic evacuation and flight up grass stems, up cacti, up sticks, anything. Those feathers came up, too. Under stinging fire they were unstuffed and unjimmied, and then the Novos hoisted the feathers and ran up grass stems and held on and hunkered, and waited it out.

It doesn't add up.

Brood adds up. "Save Thy Brood" is an ant commandment. Those white rice-grain-sized larvae and pupae are their future, their work in life, their kin, their investment; rich with protein, oils, juices, that's the treasure that the army ants are attacking for, the treasure that wants defending. So why feathers?

Who knows?

Howard, anyway, does the right thing with mysteries.

"They were big feathers, too!" he says, and he throws up his hands, and laughs.

I'm here again, looking for the dark columns on the march. The rains have washed trouble into the world in the form of dirt and gravel that has spread in fans over the desert slope, and all the *Pheidoles* and Novos are digging out, on round-the-clock watches, bringing crumb after crumb of soil up and out of their washed-in nests; but no armies are on the march around here. No one's on the move out here but me.

The feeling creeps over me again: I feel odd doing this. Why not? I am a one, one item, one scout out here peering ahead and down. I imagine myself confronted by a rancher's son, booted and jeaned, firmly hatted even in the dark, asking me what the heck I think I'm doing two miles inside his range at two A.M. with a miner's headlamp and a pack full of little glass vials, a paintbrush, and a peanut butter sandwich.

"I'm looking for army ants," I'd say.

I hope he'd laugh, too.

It may look as if I'm alone out here but I know I'm not. No more than they are, those army ant scouts, who lay a chemical trail behind themselves to find their way home, to lead others out; Tom Sawyer's kite-string in the dark.

The scouts' trail is a durable trail. The chemistry of excitement, of recruitment to raid — the chemical that sends the message "I've found food" back along the scout lines and brings a boil of thousands of stinging attackers to the victims' nest, in minutes — that message is frail, evaporates; but the home trail stays.

I follow something, out here. I'm joined up, even here, more bound than I look. I've listened to Howard and I've read his papers and books, I lay my kite-string as I go; I carry the ghosts of knowledge and desire.

I follow something. A rancher made this truck track, a double-rut like the trail made by covered wagons years ago and through strange country. Someone wrote the books I've read and they've changed my footsteps, all of them. I'm led on, warned off, swayed, and bent. How about the librarian at Yale who rooted up that stuff about honey ants? I carry message and question along. The pilot who flew me to Tucson, Pat who spread the peanut butter, and the bakery men who baked the bread for my sandwich . . .? On and on. Appearances to the contrary, I'm in company.

This isn't the same as saying I'm comfortable. Out in this subtly slanting space of basin ground with lightning jiggetting in the clouds over there, with the clouds higher than mountains, with the mountains stacked like stone libraries of time and trouble, there are times when I feel that I have been launched away. At night another planet rises at my feet. A planet no farther off than the next booted step.

Look at this: a millipede lining it out across the track, glossy as chocolate cream. He's six inches long and rounded at both ends equally and segmented in between like a zipper, with all those legs sweeping him along in sine-wave synch like cilia along a paramecium. You can see the nerve pulse run his length like a beaching wave: ripple ripple ripple and steady as a drawn line there he goes. Stir him with a stick and he hugs himself — blip! — into a spiral. From between each armored segment a spot of damp seeps: cyanide. Cyanide! What kind of world is this? No wonder I'm wearing boots.

And this: a walkingstick. He's moving over the ground very carefully, one leg at a time. So would I, if I had legs like glass threads and a three-inch-long body like a filament of bamboo.

He either feels my footsteps or sees the light; anyway he stops and begins a jerky two-step boogie, a slow rheumatic rock as if he were dancing to "Sweet Georgia Brown" at a country club do. This is his imitation of a swaying twig, and it's a good act, at least in a bush. I've seen it there and it's convincing, when there are six or eight of them all pretending that they're mesquite twigs in a gentle breeze, that a scratchy old tune has stirred their dancing feet; but it looks a little odd in the middle of the track.

And this too: tarantulas. They're common after dark here and easy to find on the sides of roads, and after I'd come across the first six or eight I stopped being kind of scared and started being just excited, though I haven't reached the point of letting them crawl onto my hand and up my arm as some people have. Some people make pets of them, and they make nice pets. They get used to you as lord and master, they don't scoot scarily but move tap tap and slowly, and they're beautiful, like articulated black velvet. They live a long time, longer than a cat or a dog. They don't even mature until they're ten years old and the female can live for ten or fifteen years after that, building white cushions for her broods and guarding the hatchlings in her burrow until they've molted once and are ready to go.

It's the males I feel sorry for, though "sorry" is out of place here. It isn't an emotion that insects know. Male tarantulas are what I find out here, and their leg-stretch is saucer-sized but their bodies are small, and they're black, not the muskratty brown of the females, and they're on the move.

They're out looking for females. If they find one and do a courtship dance that meets with approval, they'll mate; and they won't live long after that. If the female doesn't eat them, they'll die anyway. Once they've danced, their work in life is done, and there's nothing else to live for, and that's that.

The centipedes are what I don't like. I just don't like them. They wriggle-scoot at shocking speed and can bite like bejeezus, and they have all those long legs and here they come and up a pantleg as like as not; so I scoot too.

The vinegaroons, though, are just fine. They're big and black and have pincers like a scorpion does, and they can be really big like a crayfish, and they have this bulbous back end with a whip a couple of inches long at the end of that. I suppose they can nip you with those pincers if you're foolish enough to prod and poke, but they're slow to annoy. If you do stir them up they'll let go with a squirt of acetic acid which smells like boiling pickles. This is enough to make most predators wrinkle their noses and scram.

In the daytime the vinegaroons hide. They're easy enough to find, under sticks or stones in a dry wash, under stray leaves on the cabin steps. One fellow at the station caught one and put it in a terrarium outside the dining room. He put in some sticks and leaves and stones to make a nice setting, and the interesting part was that the vinegaroon moved these things around. He rearranged things, even the stones, dragging them until he'd made a hut with a leafy roof. You could see him moving around in there. He was at home.

In the daytime all these creatures disappear. At night they're out and about. Scorpions are out, too, and I did see them, and the big hairy desert ones aren't dangerous, although the tiny honey-pale ones are. In summertime you just don't lay your sleeping bag down plumb on the desert floor.

I had the feeling, looking at all these goings-on, that I was watching a kind of television. With the sound turned off. I kept thinking of some friends of mine who are very into high tech. Long years ago when it was still a ten-days' wonder they had a remote control for their TV, and the remote control had a "mute" button on it. They called this the "blab-off." They'd hit that blab-off every time a commercial came on.

When you watch a commercial that way it's very peculiar, very fast; cars zip, aspirin boxes appear in outer space, there's a soapy armpit, and so on; none of it makes the slightest sense.

I felt that way about all the activity on the desert floor. It wasn't just the silence that went along with all this scoot, hunt, sting, grab, chew, dig, forage, mate, shift, and scoot again. It was as though all of this was going on in a world from which I was removed in some fundamental way.

Which is true. Howard explained this when I asked him what a psychologist was doing studying army ants.

Animals who live in groups have certain requirements, he said. Communication, division of labor, coordination. All of the above are interrelated. Primates have solved these problems by being visual and verbal; ants and other arthropods have solved the same problems by developing a vocabulary of touches, smells. If I want to understand ants, then, there is a basic problem of language. The blab-off is a permanent condition.

Ants have been organized in cooperative working colonies for more than thirty million years. It's easy enough to see that they are organized; the question is *how*.

Howard says that most of it is done with chemistry. Each colony may have a vocabulary of some dozen or so chemicals. (We've got 26 letters in our alphabet, ten numerals, and a few operational signs, and how much can we make clear with those?) Anyway, the ants' chemicals are more than language. They're like the hormones in our bloodstream; master catalysts of shape and behavior.

These chemicals are called pheromones. They have bewitching power.

Howard is deciphering the vocabulary, the syntax, the sources and substances of this organizing language. I've watched him; it's an exhausting process.

For instance: when the slave-raiding ants go out on a raid their virgin queens go along. Suddenly, in mid-march, the virgins spread their wings and dozens of winged male slave-raiding ants appear in the air overhead; they descend, mate, struggle, die. This happens in seconds. She releases a pheromone and *there they are*.

Days of ant capturing and ant dissection and experiment have yielded this conclusion: the slave-raiding virgin queen produces this pheromone from her head. From mandibular glands, perhaps? More deciphering will tell. Anyway, a single squirt does the job; she won't ever need to mate again, not in her lifetime. Her pheromone is a commandment, a coded chemical of frightening precision; tossed, once, into the air.

Meanwhile, where do the winged males come from? Where are they hanging out now? Bark chinks, grass stems, pebbles . . . everywhere. Waiting for that single silent shriek that says it all.

Compared with them, I am a little ill equipped out here. How can I understand this ants' nest I'm watching now? This might as well be an outer planet of Regulus I'm watching.

I can see surfaces only, but even these are marvelous. This colony of honey ants, for instance; they are nocturnal, too, and their nest entrance is like a miniature volcanic cone. They're nice ants, at least to me; they're mild-mannered and honey colored. They store seed in their nests. They store plant juices and honey in the live vatlike bodies of their "repletes." When army ants attack their nest, they swarm to counterattack, fastening their jaws on the invaders, crushing and tearing, bending double to squirt corrosive formic acid into the enemy's wounds.

The daytime harvesters, the Pogos, are fierce defenders, too. They have heavy red armor, sharp jaws, quick tempers, and vastly superior size; army ants do not attack them often, and rarely with success.

I watch the Pogos, too. Most of their colony is down below after dusk but there are always a few milling at the nest entrance, on guard all night long.

I go on again, checking my sticks, eyeing the track for the sign of fifty-foot-long columns on the march. Most of all I watch. I peer at the ant world in particular and the insect world in general down the wrong end of a telescope; but if you peer through even the wrong end long enough it becomes the whole world, the entire here and now.

I never found any army ants. Something strange happened, though, on my last night of walking and stick planting and stick checking and watching. I was looking at Novos again, I think; I don't remember.

I saw it at first out of the corner of my eye, and froze. It was a very strange thing. I'd never seen anything like it in my life. I thought at first it was a ruined football that had come down the wash in yesterday's deluge. It was brown, scarred, wrinkled . . . something about it looked alive. A piece of meat? A stray chunk of cow? No, it had a bit of greenish stringy stuff, hanging. The Novos (or whatever they were) thought it was queer, too. They were exploring it and one had attached herself with her jaws and was holding on.

Then, suddenly, it *moved*.

I jumped a mile.

It did, too.

It was my boot. My boot with me in it.

Before that dawn I drove over the mountains and watched the sun rise from the peaks, and went on into the next valley and on toward Tucson. I'd packed my bags and said good-bye to everyone the evening before and the moment of leaving was anticlimactic; I came up from the desert, threw my pack in the car, took off my boots, and drove away. By morning I was in other country. It was strange to be out in the day. It was bleak, it was pale and hot; I still didn't like it much. Even after I found a tortoise in the road, which was quite a find.

With a sense of discovery as blinding as the recognition of my own boot (I've had those boots for nigh on twenty years. You'd be scarred too. They don't make 'em like they used to), I realized that I was at home here, but only in the night.

At noon I felt as gritty and exhausted as I should have felt at two A.M. No wonder, I thought, that Tucson is all canned cool, all dim interiors, swatches of artificial dark. With tubular fluorescent moons. No wonder.

NORTH DAKOTA   December

In our boundless pasturing, we went gathering up the thirst and
the hunger, the pleas and the sobs, the dung of suffering and the
timid shoots of hope, the sighs of love and the lacerated words
of pain, and we prepared a bouquet the color of blood to adorn
night's mantle.

— Jorge Amado, *Shepherds of the Night*

Shadows on the Wall

I woke up in the blue of dawn and wondered where I was. At first I knew nothing except that the windows were not in their usual places and that the smells in the room were wrong. Then I heard it again: the high, shrilling wail that had stirred my dreams and that was dreamlike in itself; it made me shiver as if I were cold. Only then did I remember that I was the new ranch hand on Mr. Wheelwright's outfit in Nevada, and that business here started early, and that the racket must mean that it was time to put my boots on and get to work.

That was more than fifteen years ago. I remember the first panic of not being able to find the light switches or any people awake anywhere. I found Mr. Wheelwright at last, only because I saw his car parked in the field beyond the corrals, and I could see that he was in it.

The sky was pink, the sagebrush was a misty humpy silver sea, and I ran out there and got in the car with him. The heater was running and he was on the telephone. He had one telephone in his car, another in the barn, and three in the house; he was famous even then for being a phonaholic, which may have been the reason for his success. This time he was chatting with a cattle baron in England. He managed to pass me the coffee thermos and then point to the field, and grin, all the while talking price per pound and negotiating a mutually agreeable transfer of frozen sperm, and I looked at the field, and gasped.

The field was full of coyotes. I'd run right through them the way you'd run through a field of yellow bees, mistaking them for flowers. They were the colors of shadow, of dawn-colored brush. They raced around the car. I could never see how many there were — three, a dozen? It seemed to me that they were playing tag and that the car was part of the game. The car was an obstacle to hide behind and jump out from. They were using it just as I used my mother or a tree when I was a child. Sometimes one would leap in the air and snap her jaws. Once in a while one would stop, and yip, his mouth open to the zenith and his chopped rippling yammer coming like a spookhouse wail, his legs taut and spread like a show dog's. When the sun lit the slopes, turning the mountains the color of milk, the coyotes evaporated like the mist.

Years later, when I had my own farm in northern Vermont, I came up from the barn one evening to find a stranger in my yard. He was wearing earphones and held an antenna in his hand. He put his finger to his lips to hush me up before I got started. He wore a black gadget around his waist and managed to twiddle a dial there and scribble in a notebook with the same one hand, the notebook braced against the gadget. His eyes shifted and focused nowhere, like a radio operator in a war movie struggling for a fix on an enemy sub. When he was done, he took his gadgetry off and put it on the ground, transforming himself into a biology grad student whom I happened to know.

He said that he had put radio collars on two coyotes and was tracking them over the country. He told me that coyotes are nocturnal and won't allow themselves to be seen unless they want to be, so that nobody knows what they really do; that one of his collared animals had been hand-raised, the wild ones were just too hard to catch; that this hand-raised animal kept turning up on people's porches and scratching at the doors and whining and scaring the neighbors half to death; and this was great because it gave him an opportunity to explain to everyone that coyotes were a harmless and beneficial part of the local ecosystem. He told me all of this before I could get a question in edgewise.

"Coyotes?" I said, when he stopped for a breath, thinking of my barn full of ewes and lambs.

"Oh, yeah. Woods 're full of 'em. It's great! 'Bye now."

The next coyote that came into my life was dead in a snare. A friend of mine ran sheep in the Champlain valley and he'd snared the coyote himself, in a hole in his fence, but not before the coyote had killed a dozen lambs and eaten the livers and left the rest. That liver-eating business is a coyote signature, he'd found that out. He'd also found out that the local legislature did not believe that there were any such things as coyotes in the state of Vermont, in spite of the grad student's surety years before.

My friend had called me in as a witness. We took pictures. The temptation to haul the carcass itself up before the disbelieving legislators was, at the time, great. We were not believed, and the photographs were not, exactly, believed either. It wasn't until a pair of coyotes had been seen, in the barn lights of a large dairy farm, just after milking time that next autumn, pulling down and killing two registered Jersey calves, that the lawmakers admitted that there were new creatures in their country after all.

It was during that time that two pamphlets appeared in my mailbox on the same day. One was from a well-known wildlife conservation organization, asking for funds to prevent the cruel and unjust persecution of the coyote. The other was from a national sheep-raising organization, asking for emergency funds to help sheep ranchers and farmers who were in danger of going out of business because of the coyote. It occurred to me that there was something very wrong in being on both those mailing lists at once.

I began to think, then, that the coyote — like most predators, and like our own predatory nature — had been politicized beyond recall. Shadows: projections of truth, highly charged archetypes of angel or demon; that's all I found out about the coyote anywhere. And the animal?

Late that autumn I saw a coyote in my meadow, only because the lights were out in the house and the moon silvered the skin of snow outside, revealing him as shadow. I knew him right away, he was too quick to be a dog, and too intent. It's the difference, I remember thinking, the difference in physical presence and grace, between a ballerina and someone who isn't.

His life is mixed, inextricably, with darkness. He's the Navaho trickster up to his old business; she's the Aztec Coyolxauhqui, goddess of the moon; he's the Spanish Don Coyote, heartless and reckless, cynical, lawless, sly, mendacious, clever. He's an animal that is always more elusive than his controversy. He had me watching, as sensitized as others are and have been; he had me laughing at things that were not at all funny, though the humor in him had more to do with human nature than his own. What, exactly, was his own?

In the end, I went west again to find out.

Coming in to Land

I'm going to the North Dakota badlands to learn about predators. I want coyotes, but I'd like to know about bobcats, too, though I'll take anything that comes: red fox, wolves, mountain lions, mink.

The prairie rolls now, slowly, under the belly of the plane. Its colors are dun, tan, like fur. We're going north.

The grasses of the prairie lie smooth as cloth, rolled like hide over muscle. North to the Canadian forest and west to the Rockies and east to the Mississippi and south to Kansas the plains make up one — one varied but one — space. From here, its unity is clear even under the wintry farm-field quilting, the etch-lines of roads, the target-shaped circles where irrigation machineries crawled round and round like mills.

As we go on north the fields thin, scatter, fade. The land dries and chills. Snow blurs the dun grass.

I turn back to my notes, my lists.

What do I know? Not much. Nowhere near enough.

I know that most predators hunt at night most of the time, though they make more use of dawn/dusk hours or of moonlit nights than of deep dark, and some are more flexible in hunting time than others. They hunt when it works best to hunt. It just happens that most of the time the night works best, for them.

They have predators of their own, and prey. They have territory. They have language. I make a list of more things now, more questions written out, adding to the ones I've written out before.

The sun is low; the light gilds the flanks of hillocks down there as though they'd been plated, there's a gray wash of brush in hollows. In the distance is a line of darker hills. It all looks watercolored, inked, with the abstract subtleties of wild ground. I see no roads. The farms have vanished. The snow lies thinly everywhere. I look down at my feet, in their stockings and little shoes, and they look wrong, anachronistic, strange. I notice that the feet next to mine are wearing boots. They are boots of tooled and stitched leather with pointed toes, stacked and slanted heels. I know something about these boots. These are the fancy pair, the new pair, the city pair. The good old pair — with popped soles, rubbed toes, downed heels — are lined neatly on a porch somewhere, under the jeans jacket with the wool collar and the hat with earflaps, waiting for this man.

"Excuse me, ma'am," says the wearer of the boots. "Can I buy you a drink?"

I look up past the snapped shirt and the black leather jacket to a lined face with dark hair slicked neatly back.

"Thank you," I say. "I'd like that."

"Whiskey?" He winks.

"Yes, sir."

"Whiskey," he says to the stewardess. "Make that two."

I'd come west.

Not that Arizona wasn't as far west. When the Apaches still rode them those desert basins were the last southern fingerings of these same plains. When the first settlers came there the basins were seas of grama grass. That green-gold perennial grama has been grazed away by cattle and competed out of existence by greedy mesquite in less than a century.

"I run cattle," my seatmate says, in answer to my question. "I run my cow calf pairs on native pasture. Buffalo grass, blue grama, western wheat. If you overgraze then the brush comes in and you have to let it rest. The only way to treat native land is to let it rest."

We are into our second whiskey and I'm getting the picture. He's a badlands cattle rancher and he's setting the scene: our respective destinations are tens of miles from each other and here that isn't much. Outside the window the plains are ashen with dusk, the sky rouged with winter cloud; the sun is gone.

I know that my search for night life is a search through extremes: south to north, east to west, hot to cold, summer to winter, and more than that: there are extremes, too, in the sources I've chosen to use. It occurred to me months ago that the people who would know most about nocturnal predators would be the people who have the most to do with them. The people who have the most to do with coyotes and bobcats (and foxes and mink) are predators themselves: trappers, hunters.

"I figured they would know more about the animals than anyone else," I say, in reply to my seatmate's question.

"You're right there," he says. "So why here?"

"I looked up fur-trapping data. You know, where the most furs are harvested. North Dakota looked pretty good."

"Good! Heck, they've got the bobcat down to where a fella can run a ranch, but the coyotes! They got it all wrong down at the county courthouse. We know who owns the country." He laughs, pounds the armrest, shakes his head. "After dark they're just about trippin' over each other to get in the barn door. Now they're talkin'" — he takes a sip of his drink, rattles the cubes, looks at me out of the sides of his eyes — "folks from out east mostly, pardon me, ma'am, been tryin' to tell us how to run our wildlife out here. Talkin' about bringing the wolf back into the country."

"What do you think about that?" I ask.

"Think!" he says. "I *know!*"

We laugh.

The timbre of the plane's engine changes, I chew an ice cube to pop my ears, I feel the thunk of landing gear. Outside the window is a uniform velvet darkness. A brief splash of city twinkle is set in the dark, with lit tendrils, like a ganglion. Bismarck, North Dakota.

"Good luck with the wolves!" I toast my seatmate with the last of my melting ice.

"Good luck to you," he toasts me back. "If you get them fellas to talk you'll have something!"

"I hope so."

"I *know* so."

Predators

I remember . . . well, the truth is, I remember not much! I remember the caterwauling of breeding bobcats one night in late winter, when I lived in the mountains, and how it made me huddle in my bedclothes and go out into the next morning with my eyes wider than they had been, so that every clack of a winter twig and rustle of dry leaf made me freeze and look for the source of the sound. I'd caught something of them. Something of my own, perhaps, from them. The day was changed. No sound, no taste of wind, no shift of shadow, was without a meaning of its own.

Now I feel this again. I wake up in the darkness of a room. I lie still, waiting for my eyes, ears, nose to tell me what's up.

I've been in North Dakota for ten days now and I've learned more here than I can absorb with any ease. What do I really know? Still not much. We are born into mystery, we inhale it with every breath. The more we know the wider the mystery goes, like a landscape opening when you lift your eyes. The predators are everywhere. Look at a range map of coyotes or bobcats: there they are.

Almost everyone has walked through land claimed by them and hunted by them. I am sure now — surer than I am of most things — that they have sniffed the scent of your track, too, when the dew fell and the shadows gave them cover to cross your trail. They know you; more about you, perhaps, than you wish to have anyone know. You may have walked that place a hundred times with your eyes open and not seen signs of them; but this means nothing except that they have practiced their profession well.

I am beginning to understand what this profession is. Lozenges of light resolve themselves into windows, the humps of shed clothing, the gleam of mirror. I'm in a room in a house near the Killdeer Mountains; the lit hands of my clock tell me that my old insomnia has rousted me out; an insomnia based on old inbuilt rhythm. All night our sleep runs in seventy- to ninety-minute cycles, deep to shallow and back to restful deep again, and in the shallows we can wake, easily, to take stock of our surroundings. We are our own town criers, it turns out. My senses tell me here and now that all's well. I'm safe here. I could sleep again, but I will not.

Making the lives of bobcats and coyotes into some coherent shape has been like reading stubborn runes. These two are close cousins of domestic cats and dogs, so that much of their behavior should be something I can intuitively understand. Some of it is. Domesticity is difficult to see through, an opacity of the spirit, but I need the head start. I haven't expected to see even one coyote, even here, though they're common enough. I can see them in zoos, of course, if I care to, anyone can, but that would be like watching a jailed felon and drawing conclusions about the behavior of man. An animal that is captured is an animal that has failed. An animal that is watched and is unaware of the watching is an animal that may be captured. Contrary to popular myth, it takes more than a dose of luck to see an animal and watch it be in the wild.

Wild: what does that mean, anyway, except that they are as independent, as adept, as central to nature as we think ourselves to be?

So. I have had to learn by second hand, by allying myself with a teacher or two. This is what I've done.

I have had several teachers here. There was Steve Allen, a wild-life biologist in Bismarck. He's originally from Missouri, his granddad homesteaded for five years in central Montana, he's still more at home in a wool toque and jeans and boots than office clothes. He's a fine biologist. His research on red foxes and coyotes has answered a host of questions about these animals' relationships with the land, with each other, and with their prey.

Then there was Cobby, a fur trapper from Idaho. And Bill Austin, from Wyoming. Years ago Bill was a coyote bounty hunter; nowadays he's a self-styled coyote researcher, one of the best.

Now there's Gary Jepson. Gary has consented to more than an interview. Weeks ago he invited me to live here in his house, with his family. He took me on in the spirit that I need to be taken. He is, by all reports, a master trapper. He has treated me as an apprentice to his trade.

He is a quiet man, with a wary reticence about him that extends his magnetism, just as his black Stetson extends his already considerable height. He teaches in the tradition of good teachers, which is more by asking than by telling, though he does tell. He shows more than he tells. There is a good deal, I know this already, that he won't tell. There is a lot that he'll let me learn for myself.

Today as other days I will tag along when Gary runs his trapline. I will ask questions — though I've learned not to ask too much. Silence teaches, too. The ground teaches. I'm getting used to this. I learn by watching him.

I do have that neat list of questions that I made back home and amended in flight, a list made after doing a lot of reading — a lot of reading, mind you — but that list seems strange to me now. It seems, in more ways than one, to miss the essential, to strike only glancing blows. All the time I find myself (silently often; aloud sometimes) asking "Why? Why? How?" like a three-year-old.

I get up now in the dark and get dressed, tucking and buttoning by feel. I don't want to wake anyone up. I want to go outside alone.

I go down the hall quietly and past the table and chairs and through the kitchen to get at my boots. I've brought the wrong boots; the felt-lined jobs seemed too big to pack, clumsy enormities as they are, so I brought these hiking ones. I've resigned myself to cold toes as a result, cold toes being something my years of farming in northern Vermont have taught me to respect, and even bear, but I don't like the experience any more than I ever did. Less, in fact.

Once my boots are laced, I stretch, grunting twice. I feel for my coat, fumble on the shelf of hats for my hat, and go out the door and down the steps.

I might have been doing this every day of my life. I feel as though I have.

The winter air is clean and sharp like a plunge into river water. Above the shapes of house and barn and the feathery line of shelterbelt trees Killdeer Mountain rises pale in the sky, smooth-crested and bright, naked in its moonlit snow.

I've come outside to get my bearings. This has been an experience so alien that after three days I feel as though I have never done anything else.

Out here now there is nothing to see but the moon and the stars. The lower sky has a pale opacity that hasn't collected itself yet toward the east or anywhere. It's the time of day that a Utah sheepherder I once worked for called "the crack o' day." Celestial twilight is another name; more magnificent, less clear.

What is clear here is this: predators are animals of power. I observe this now in the dispassionate way that one does when one's world has been turned on its ear. The constellations of a winter morning are upside down; I have to tilt my whole self to recognize a single thing, to see that the stars have not changed, too; that the two Ursas clamber around the North Star just as they do at home, and Draco the dragon looks, as always, as though he were teasing the bears, or as though they were all three playing tag. I find Leo by drawing him with dot-to-dot and much filling in around the bright star of Regulus. He's at the zenith, lording it over the rest.

I hear a stamp and a snort and smell horse all at once. Warm horse-breath, heavier than frost-dry air, drifts through the yard like a river. This is what happens at night: warm animal smells fall to the ground.

"Hey," I say. I go over and stand against the pasture gate. Horses are more prick-eared in the dark, ready to stamp and bolt. They are always alert for any sudden thing, for the flash of the sneaking strange, and a white stone or a piece of blowing paper will make them shy and jump off sideways as if these things were a snap of jaws or a clawed leap for the throat. They're about to shy off now. They need to know what it is that I am, so I stand at the gate where they can get to me, where they can discover that I'm strange maybe but quite unpredatory. Once there, I go on looking at the stars. Pollux, Castor. Arcturus. And next to Arcturus the little Corona, the crown, a pretty cluster. One horse puts its nose on my neck and blows like a furnace, and wibbles a bit of my hair.

I wait. It takes fifteen minutes for my eyes to adjust to the night. Half an hour does a better job. A flashlight shone, a house light glimpsed, anything will bollux the change and then one has to start again. The pupil dilates. The nerve networks that process sight signals increase their sensitivity up to a millionfold. The retinal cone cells, which give color vision, don't work in low light. The cones turn off; the rod cells turn on. (These things are named after their shapes; I suppose this is obvious.

Anyway, the rods and cones are light-sensitive cells that line the retina and are attached to the nerve fibers that shunt information in.) The rod cells are filled with purple pigment. This pigment bleaches in bright light, and its rebuilding takes time. Even when this pigment is up and running and the rods are working at their best, they don't give the sharpest messages, and most of them are clustered at our retinal fringe, so that part of good night sight means becoming aware of what one sees, and can only see, out of the corners of one's eyes.

The human eye is a wonderful thing. Decrease the light level one-hundred-thousand-fold, say the books, and give your eyes half an hour to fool around, and you'll see things as well as you ever did. In black and white maybe, and fuzzily, but well enough.

There is the seen and there is the unseen; my fingers explore the cool metal of the gate, the crevassed wood of the gatepost, the line of wire rippling like a body curve and stippled with its wrapped barbs; the mind is more marvelous than any eye. I remember once holding in my hand an address book that belonged to a blind woman. It was filled with thick white pages. The pages were blank, dimpled with constellations of bumps. There was nothing in it that I could read. The mind hauls information from the world all the time, through every pore; seeing well in the dark means knowing that you can never see much, that there are other ways to know the world, that these ways can be used.

Now I can see as well as I ever will. In the light of a half-moon the new ranch house looms low against the slim windbreak of trees. There are pickups in the yard, and the barn with — faintly — the dark shapes of chains and traps hanging in rows against the gray-white wall. The old house, closer by, is dark and square. Since the new house was built, this one has been turned over to the fur trade. I've been inside. I've worked there. The old living room is full of wooden stretchers of all sizes, some empty and some in use, for coyote, fox, mink, muskrat. There are hoops for beaver. Dried and finished furs, brushed thick, hang in coat-hangered rows in one of the bedrooms. On nails along the old kitchen wall are shriveled things the size of figs and tiny filled sacks, and on the counters are neat rows of kettles, bottles, and jars, some full, some empty. The kitchen looks like an alchemist's place. Which, in a way, it is.

It's the work of an alchemist to probe the unseeable, to elicit poofs of response from liquids tossed together, heated, shaken; to reveal the haunting sucking force of metals; to capture in some corked vial the catalyst, the directed potion.

To have the mind of an alchemist is to know that there are forces that change life, and the earth, too, and the universe at the core of which is, always, one's own heart.

I don't move for a long time. The prairie goes out from me forever, and the sky is too big to leave.

Then I cock a leg over the fence and am over among the horses. They make way. I move over the ground to a slough and across again to where the ground rises toward the sky. The horses follow me; one tosses her head and I see the whites of her eyes; her forefeet come up off the ground.

"Hooooo, hey," I say. "Hey."

She comes down on four feet again and stops, and they all stop, and congeal to dark shapes. They stare as I go over the ridge.

When did I learn to talk to horses? I don't remember. It's all in the tone of voice, in the fearlessness, in the showing of respect. It seems, at the moment, a good thing to know.

The country moves under me as if I were walking on frozen waves. The grass crisps under my boots, my foot sounds muffled by an inch of snow. From the high ground the country seems illegible like a scroll of Arabic script that unrolls, infinitely, the size of the sky. The snow makes this much clear even at night, as if the ciphers were backed with white paper or cloth.

I am the most visible creature here, the most blind and unaware. I feel the prairie emptying around me as I go: I trail the signals of my self out here and everything but horses knows to stay away.

There's only the shush of air moving past my ears. The land flows the way water does, moon-slicked, coal-dark. I have never been anywhere with so much air. The flow of the ground seems to have no direction in particular and has nothing in it to say which is height or depth, so that I feel as dizzy as if this were an ocean halted in mid-heave that could un-halt any moment, throwing me loose; throwing everything loose. I'm glad of the unsteadiness; my feet know solidity even when my mind refuses to oblige; there's an excitement to this like that first toppling dive of a roller coaster, or a Ferris wheel's going over the top. Even the fields and the plowed ground seem flung anyhow over the land's supple roll, and the perspective of the fields is vertiginous, like one of those clever drawings made to fool the eye. Snow fills the etch-lines of harrowed soil so that here runs a swirl of striae, there a spotty swatch of sunflower stubble, then a rough meadow or a stitching of shelterbelt; it looks frail, unsteady.

A person, here, is small. The scant lines of roads go out squared across the country, running on section lines. Here and there are ranches, nestled in trees, as if in the palms of hands; otherwise the country rolls out as bare and clean as it ever was. Overlaid on the man-map of section lines, houses, roads are maps of another kind.

Now in the brushy hollows of the country the white-tail jackrabbits — prairie hares, the hares of seventeen-foot leaps and forty-mile-an-hour sprints — feed, hopping, nibbling, on twigs and buds. Their winter range isn't large; a few acres of brushy forage. They have daytime hiding places where they can lie, still, concealed, the color of the ground. The cottontails forage, too, at dawn and dusk, rising to their haunches to nibble high, or digging ratlike for snow-protected grass. The white-tail deer clip buds in the aspen thickets, wrenching twigs with an uppercut slice of their heads; now in the moonlight they're clustered in winter herds that have gone the gray of bark and shadow. Overhead the sharptail grouse and turkeys roost in the branches, humped and fluffed, tails drooping, waiting for dawn. Underfoot the red-backed voles and deermice scuttle through their quarter- and half-acre grounds, keeping to their runways and trails, foraging for roots, grass leaves, seed. Badgers come down the hollows of their own range, snuffling, digging, hunting the voles and mice.

Now in the long grass of the damp sloughs the meadow voles forage along their systems of wet tunnels. They gather the leaves of grasses, the buds of buckbrush. Longtailed weasels patrol the sloughs with a leap and a leap and a liquid slide, hunting the voles. A raccoon comes, too, a humped shadow among shadows, smelling out roots, garnering the corn and wheat and sunflower seed that the harvest has left on the ground. Muskrats chew roots of cattails and grass and come back to their cattail huts to comb their fur.

Up on the grassy plains the gophers' burrows pock the soil, though some are inhabited, too, by the whistling grasshopper mice who are foraging now, in the dark, hunting for cold-stunned insects, tucked cocoons, even other mice.

Maps here are overlaid on maps; every three miles, more or less, another family of coyotes claims the whole ground. They patrol their boundaries and hunt there, taking their own trails and offshoots of trails, circling the brushy draws and shelterbelts, eating frost-sweet berries, insects, small mammals, hare and cottontail, grouse and turkey, dead cattle and trash, whitetail deer, pronghorn antelope, cats and dogs; whatever they want at the time. The territories of the red fox are smaller, and are tucked in the interstices of the coyotes' claims — in what to coyotes makes for less desirable hunting — roads, farmyards.

The bobcats hunt a different kind of ground. They have solitary claims, and hunt alone. They hold the rocky badlands by the rivers, the forested mountain slopes, the rimrock, the high country. Now their eyes are black, their pupils — once slitted

Raccoon
Procyon lotor

together like hands folded for prayer — are opened to the full night form of a globe.

All of these hunters have, as their first principle of work, to decipher the maps and signs of their prey. Human hunters have to decipher the maps of these animals that are their own prey, and they need to learn to read the languages in which the maps are written. A few go farther than this, because it takes the better part of a lifetime and more discipline than most people have, but Gary is one of these. So is Bill Austin. These men are learning to speak the predators' languages themselves.

This is what I am beginning to understand. These maps, these idioms. They are strong, clear things. They are here in the dark; before us and after us, and in spite of us, these animals go on laying out their ground and their lives in song, in scent, in code: *This is mine, this is what I am.*

I go on over the ground now, sensing fences before I come to them and going through the strands of wire and on again. The air is cold and clean and tastes of stone and snow. Now and then I catch a warmer whiff of hay, manure, animal. If I could read this as they do! If I could taste meaning in the air!

I am resigned to patching pieces, shards, anecdotes, images; to assembling night animals from fragments, like a collage. I work here within the limits of my own senses, my own language. Too often I have to read from daylight clues what it was that happened in the dark. I have to read signs backward to what was.

I have to take a plunge; to join what it is I'm after; to trust things from the depths of instincts that have been dulled by domesticity, glossed by culture.

Given that I am alive, and animal, and human, this is difficult to do.

The first night that I was here Gary told me about the coyote wars, the bounty-hunting days, the days of the now-outlawed cyanide guns and 1080 poison; the wars that failed for any number of reasons. Most of all they failed in their goal because coyotes knew more or learned faster than any man.

"Any way you tried to go after them," he said, "some of them would figure it out. This is the point I'm trying to make: they learn. They communicate. They're instantly aware of anything new." He drove home his points like nails, one after the other, looking me straight in the eye; this was catechism. "They're curious. They'll try anything, once. They learn from one another. I assume they teach their prey in the same way we teach them. They're continually . . ." He paused; ". . . those that survive are continually getting an education."

"I understand about the education," I said, "and all that. But it doesn't seem right to me. To try to wipe them out, I mean. Those were government programs, I know, the coyote wars and the buffalo slaughters and all that. I don't like the attitude! There was no respect or concern for the animal. For his *life*. For his place in the scheme of things. For his beauty!" I said all that. It was what I believed, so I said it: "The buffalo are gone, the elk are almost gone, the wolves and the grizzlies are gone, the kit fox is all but gone. I was brought up to believe that wildlife slaughter was morally wrong."

He was quiet for minutes before he replied.

"I hope you're not thinking that's what I do," he said at last. "You wouldn't be here if you did."

"I'm here to find out about the animals. I'm not here to find out about killing them," I said.

"It's not trapping that endangers wildlife. It's development. Pollution. You know that."

"Yes, but —"

"Listen. Let me tell you this," he said, his voice low. "If you trap a certain species, you catch him and skin him, you come to realize that you're dealing with creatures that are highly spectacular. You come to get attached to that animal. It's more a mutual respect thing. I've never known a trapper trap any animal for any length of time without getting involved like that."

There was a pause.

"I'd have a hell of a time killing coyote pups," he said at last.

"I'm glad I don't have to. But I know when he's full grown he can take care of himself. He knows that, I know that. If he gets caught it's his lookout."

That was as close as we ever came to an argument.

Now I've come to an edge, to a brink in the grass, and the ground rolls down into the pale planes of cliffs and dark furrings of trees. This is an arm of the badlands: this is as far as I go. Away and down is a long glitter of water.

It's this river that pins me down, it's something to which I can peg my existence, it's finite, a bottom to the world. This tributary of the Little Missouri, full of pale moonlight and sky-shine, goes wandering in the great eroded plunge of its valley between bars and loops of trees; but even here, on the height and looking down, I'm unsure of the direction of flow; even here, there is no sure feel of *tilt*.

I turn back toward the mountain and the ranch. My feet are numb. I want coffee. It's gray-light time, almost dawn. I can see, suddenly, distance. There are specks of cattle on a far hill but nothing moves, no smoke rises, and aside from the fences there is no sign anywhere here of human habitation. I have one more day ahead of collecting pieces, shards, signs. I am more aware of sound, click, shape-shift, whiff than I ever was; but these don't tell me what I need to know. I am not aware enough.

Before the sun is over the horizon we're in the truck and off. Gary wears a checked wool deerstalker cap with the flaps unlaced, and jeans, and a brown jacket over a flannel shirt. It's only a few degrees above zero, but his jacket is open and his gloves are on the dash along with the plastic cups. Between us on the seat is a loaf of sandwiches, a sack of apples, a plastic container of cut oranges, a whole bag of candy bars, and a coffee thermos the size of a small bomb. In the back are bags of trapping gear propped among the clutter of shovels and tires and chains endemic to pickups everywhere.

Gary is more than six feet tall, and lean, and looks the part of a cowboy (which he has been, and of the old-fashioned free-lance variety, too) or a rodeo rider (which he also has been) or a rancher, which he is. He has twenty-five hundred acres here on which he runs cattle. Everything about him is out of scale: the mammoth coffee thermos, the mittens the length of my forearm.

Gary has trapped fur in the prairie every winter since he was ten. Here trapping is a common livelihood, a skill that goes back even before the days when the Dakotas were explored and settled — if you could call it settling, which it wasn't exactly, not yet, not then. Trappers were the first non-Indians in here. Trapping was what the Indians did here. It was the only thing to do. Now, for much of the winter, it still is.

The animals he's after are what I'm after: everything that he traps is active at night or at dawn and dusk, and, except for the beavers and muskrat, all of them are predators. We're after the coyotes and foxes and bobcats for different reasons, but these do not seem so different as they did before I came. I know this sounds odd; it seems odd to me.

Our first stop is just above Nard Creek. We bump along a track through a field of placid Herefords. They look as bright as stuffed toys. At dark they will bed down, together, on a rise of ground, their heads in different directions, and will surge to their feet and stamp at anything untoward, just as horses do. Their sleep is never deep. The instincts of the Aurox are not wholly gone; they haven't lost the deep knowing that manifests itself as fear, the knowing that goes to the dawn of diurnal and grazing things: night is predator time. Their instinctive stamping tells the predator that they are warned, aware, ready to run, that his chance of success is poor. In daylight they are not afraid of anything. They move off the track now with massive serenity as if they were boxcars full of beef, shunting, with a sensual sway of rumps.

We come to a set of corrals, nice corrals, all metal with red gates. My job is to be the gate opener on the team but I make all the usual mistakes. When I shut the gates behind us, I find that I've shut myself on the wrong side and I have to clamber over, both times, and though I pretend I couldn't care less and that I enjoy the clambering I haven't fooled anyone. When I get back in the truck, Gary gives me a long look from under the brim of his hat.

There's a steep track down to the creek and we slide and wallop halfway down, banged around like beans in a bucket. We park and walk and we check the whole creekbed, five traps and eight snares. All of them are empty.

The valley is thick with coyote tracks. They've run within a foot of his traps in the darkness and they've shoved the snares aside or pushed under them, though it's all I can do to see the difference between the loop of a snare and the curve of a bramble. Gary says that there are five animals using this valley and that they do most of their hunting here and that he hasn't caught a one. We are careful to walk as far from the traps as we can. The coyotes will know where we've been. He makes minute adjustments to the snares, and sets three more.

"These coyotes know, exactly, what I'm doing here," he says. "My only hope is that one of them will get careless, some night, and forget what he's doing." He looks at me again, under the brim of his hat, and there is a ghost of a smile before he looks away: "Coyotes, generally, have their minds, exactly, on what they *are* doing."

I manage to be on the right side of both the red gates on our way out. I have nothing to say for myself for at least an hour. The coyotes' world is beyond my grasp. The country unreels around me as if I were running through it myself, in the dark.

There is more than one way of knowing things and more than one way to approach the finding out. I gather as much now from the hiss of blowing snow and the sighs of the aspen thickets as from things written: this wind, after all, is in their world.

Things written have their place. Scientific papers never say enough about the expense and exhaustion and exacting labor that went into making them, or the experience gained in the making. These small pamphlets almost bury their revelations under footnotes and graphs and sober scientese, but there they are: they are the ends of Undertakings. I have read a great many papers about coyotes, and have garnered only crumbs, until last week.

Steve Allen is a scientist, and, among other things, research is what he does. More than ten years ago he began a study designed to find out if coyotes live in discrete social units in the population — "families," in other words. He wanted to know what a coyote family was, what the family territory was, and how family and territory changed over time. Information about these things was anecdotal and incomplete: not good enough. Scientific game management was and is his goal, and there were things that Steve needed to know.

"We were in the dark, see," he said. "Most ways we still are."

Steve's office was cramped by shelves full of books and filing cabinets and desks. The walls were a clean warm cream, and a cased wolfskin hung, full length, by his door. I had to sidle between wolf and bookshelves to get in.

"Everyone wants to see a wolf." He grinned. "That's closer than most of us will ever come."

The nose almost touched the ceiling tiles, the tail almost brushed the floor.

"It's bigger than I thought," I said.

"That's what most people say," he said, and laughed. "And he wasn't a big one, either."

He regrets the passing of the Dakotas wolves, and the buffalo, too. He gets a faraway look in his eyes when he talks about buffalo: the roll of dark backs and bawlings and dust; a morass of buffalo, a stench, a storm.

"They used the whole plains better than cattle ever can," he said, shaking his head. "When farming came in here they had to go. They don't jump fences too good. When they come to a fence they'll auger right through clear to Kansas," he said. "Ranchers now, they won't put up with that."

Ranchers do have to put up with coyotes, for better or worse, and coyotes have become a game animal here of no small consequence; and managing game is Steve's business. In order to do this business well, Steve launched his coyote study in 1976. Every April for three years running, all the coyotes in three hundred and sixty square kilometers were chased down by helicopter or small aircraft, immobilized with a hundred milligrams of ketamine hydrochloride shot from a dart gun, fixed with numbered eartags and radio-telemetry collars, and laid out in the shade to come to their senses.

The coyotes were radio-tracked with portable receivers and hand-held or vehicle-mounted antennae. The radio tracking went on (says his paper) for six or more consecutive hours over ninety-seven different days and two hundred fifty-two different nights.

Steve found, first of all, that each territory was claimed by a mated pair. In addition, there was a single lone male coyote whose home range was larger than any of the family territories. Unlike the families' ranges, his overlapped considerably with the territories of the others. Mated or single, these coyotes did almost all their hunting and moving at night. From April to September the families reared pups; in fall and winter the pups dispersed to find their own home grounds but the adults stayed put. When one of the adults died, its mate often mated again with a young animal coming in from somewhere else. When both adults died, their territory would be claimed by a new young pair. The size of the territories stayed fairly stable no matter who claimed them. There was some shifting of boundaries with new inhabitants, but not much.

The results of the study can be set down in eight pages; but for Steve, for those years, it was his life. I picture the truck littered with data sheets, rolled grid maps, pens, pencils, a thermos, spare mittens; with its lit dials, the pinging of its antennae, the crackle of its CB radio — the burdens of humanity: wheels, paper, engines, maps! Coffee! The night country with its rumpled winter meadows, its summer fields thick with wheat, the lighter sky and the horizon of grass, trees, the dark angle of a barn roof, the pinging . . . pinging . . .

Steve always knew where the coyotes were, and, he said, they always knew where he was. The pinging would tell him that a coyote was crossing the road just ahead of the truck but Steve could see no shadow moving there against the pale tarmac; the animal crossed through a culvert, or in a hollow, out of sight.

He knew who they were, too, just as they knew who he was. The published study labels the coyotes M7 or F3 and so on, but Steve and his research assistants gave them names. The study deals with the general; more and more as time has gone by, as time went by then, night after night, Steve came to know the particular.

There was one pair that interested him more than the others. They were named, for reasons long lost and that never mattered anyway, Crankshaft and Petunia. Crankshaft was a nine-year-old male, Petunia was a fifteen-year-old female, three-legged, a seasoned veteran of the coyote wars. They were a loyal pair, savvy, elderly. They did all of their moving and hunting in the dark. Often, they hunted separately. They had their own methods, their own tastes. Their territory included a large sheep ranch, but the sheepherder swore up and down that there wasn't a coyote in his country.

"Petunia had learned, in the coyote wars, see, that sheep were trouble," Steve said, "and she raised her pups to leave sheep alone."

One winter night, Petunia died while she was traveling through a cornfield. Her collar was still sending its signals, and after the signals had been suspiciously immobile for hours Steve went into the cornfield and found her there. He took her in for an autopsy. According to the vet's report, she died of an aneurysm at the confluence of the anterior and posterior venae cavae: a heart attack.

After Petunia died, Crankshaft left their old territory and took up a solitary residence in part of the other lone male's home range. After Petunia and Crankshaft's range had been empty for a matter of weeks, a pair of yearling coyotes moved in.

"Those youngsters didn't know to lay off Alfred's sheep," Steve said, leaning forward, elbows on knees. "The ranch went out of the sheep business in less than three years."

The coyote research team saw Petunia hunting only once. It was just at dawn, they were flying over the study area in a helicopter, and they saw her laid out on the ground as still as death, as flat as a rug. She was in the middle of a grouse dancing ground. She was the color of the prairie grass.

The male grouse were displaying there, whirling and shuffling, raising their tail feathers in their courtship dance, each on his private patch, bobbing and calling to attract a mate. The females were stalking the periphery of the dance ground in pretended indifference, making up their minds. One of them would, sooner or later, step on Petunia. The dance would end there.

"Three-legged coyotes learn to do things that are never required of a four-legged animal," Steve said.

He looked at me hard then as if to make sure I got the point.

"I'd been studying coyotes, see," Steve went on, "and Petunia had been studying grouse." He leaned back in his chair as he said this, as if there were as much satisfaction in this as in the other conclusions he'd come to. "She'd studied other things, over the years. When she lost that leg — to a trap or a gun — she paid," Steve said, "her college tuition."

Petunia lived on well into coyote old age. She raised ten pups with Crankshaft that last spring. She learned to make a living and to evade her two-legged predators, just as her kind had once learned to evade the wolves. She earned a life. She died fat.

*   *   *

Now Gary and I check the next five traps without getting out of the truck. We drive in loops and curves across the prairie where there are no roads. Two traps are along a watercourse that runs between round grassy hills, and three are in a triangle around a cattle pond. I see no sign of them, of course; no one should.

We stop by the pond and have coffee. The pond is half frozen, the water black and the ice white; in the dun bowl of grass it looks like a painting by Georgia O'Keeffe.

I find myself asking the same questions over and over:

"How would they move through here? Why?"

He answers me by pointing through the windshield:

"Here they'd come down the way the water does, here, down the draw we've come by, the easiest route. Think of water, it runs the easiest route." His eyes flicker over the landscape as though he were reading print. "The bigger the watershed, and the deeper the drainage, the more they'll use it. They'd circle the pond — most every fox or coyote would circle the pond, or at least the downwind portion, and they wouldn't strike off south because they know the fence is there. They'd go over the saddle."

I begin to have some sense of motion. Sometimes I find I can begin to read the landscape, haltingly, as if I'd half learned the alphabet in Greek.

I know that clumps of long grass left by the plow or the mowing machine may be scent posts for coyotes or foxes; that they'll be sniffed for mice and voles and other small game. A patch of brush or trees will be circled to catch the scent of deer, turkeys, hare. Water makes good hunting places, too; cattle ponds, rivers, sloughs. The bottoms of bluffs, the lines of a fence, the beginnings and ends of draws: this is where animals have to travel on their ways from sleeping place to feeding place to water. Wherever two travel ways meet — where a draw meets a river, where one fence or road meets another fence or road, where a track crosses the dam of a farm pond, the cross-roads of the country — this is where animals have to pass.

We go out now over more dry prairie and down into badlands along the drainage of Chase Creek. We park where the road looks too steep to be safe. The snow is slick under the wheels.

"I call the badlands mountains in reverse," Gary says, and grins.

That's what they are: a world of slope and crevasse and rimrock carved down into the smooth roll of the plains, so that you don't know it's here until you're on a brink. You come across the prairie and here you are.

"Badlands," he says, "is what the Indians called it."

The word carries a pidgin tang, a translator's simplicity.

"The Indians didn't come in here," Gary says. "They were afraid of it. Grizzlies denned here, wolves lay up in here. It's rough country to get around in."

This is an understatement. We walk down a path on snow as fine as dust. The slopes rise around us and shut in the sky.

"The wolves and grizzlies, two of the old top predators, are gone," Gary says. "They are not compatible with agriculture. And won't ever be."

This is simple, too.

"The lions now," he continues, "they are still fairly plentiful. Once in a while one of them develops a taste for sheep or cattle and then he has to go. I don't know how compatible they are. But they're here."

"What effect did the wolves have on coyotes?" I ask.

"Well, predators compete," Gary says.

"Are there coyotes in here, too, then?" I ask.

"Now there are," he says, and nods, and I see I've asked the right question.

"Grizzlies and wolves and mountain lions keep control of the small predators," he says. "A mountain lion will kill a bobcat if he can, a wolf will kill a coyote, if he can. A grizz will stand up to anything and anyone if they're in the way of something he wants. With the big predators gone, the smaller ones are more numerous here than they were when this country was wild."

"That's what Steve said."

"Then it's got to be true," Gary says, and laughs.

The track angles down toward the bed of the creek. Below us the ground levels and fills with brush, stunted cottonwoods. With a shift of shade and color I can see this as if it were night: the moon-silvered screes, the blankets of shadow, the cave hollows of rimrocks staring like eyes.

"I wouldn't like it here at night, Gary," I say.

"The Indians said it was haunted," he says. "I believe they were right."

The path narrows and there is a wall of stone to our left, a plunge to the right. He stops in front of me so suddenly that I almost run into him.

"Look at that!" he says. "My God, that's a big cat."

The tracks are clear in the snow; two inches wide. A house cat's would be just over an inch, a mountain lion's would be three.

Gary sighs. "My God, that's a big tom," he says.

"Will he be back?" I ask.

"Maybe in a month, six weeks. He'll be back," he says. "Toms have a hell of a range. They'll overlap with as many females as they can. They'll mate with all the females they can. They'll kill any kits they can find."

Silence descends on us then like something animate, with a force of its own. Planes of snow and stone rise, jumbled, as if they were falling all the time; pencil sketches of brush emerge from cracks, scrabble at air.

"We like to take out the big toms," Gary says softly, still looking at the tracks at his feet, clawless pad marks laid down like ink on paper. "The primary cause of cat mortality in the first three months is the toms. The females can generally protect their kits against the younger toms, see, but this fellow . . ." He shakes his head.

*   *   *

I remember what Cobby told me about bobcats.

"They're moody," Cobby said. "Moodier than a coyote. They're like house cats, only bigger."

He told me that cats have their core ranges in deep timber, rocks, rough country. They have fringe areas that they patrol less often. Most people hunt and trap only in these fringes, which is a good thing, according to Cobby. Cats reproduce more slowly than coyotes do, so their population is easier to damage, though it's rarely damaged as much as some people think; most people don't get into the country where the cats are. Cats are canny; they leave little trace of themselves anywhere. They're solitary. They're more nocturnal in summer than in winter. In winter they hunt at dusk, dawn, and even into the morning, not in deep dark. Their territories overlap with each other and interweave.

"A lot of house cats will use the same alley, right? Well bobcats will use the same canyon, the same path up the same rock," Cobby said, "and unless you know exactly what to look for, and what that is most people never learn, you'd walk that path a thousand times and never know."

"They don't talk to each other, do they?" I asked.

"Not what you'd call talking." Cobby grinned when he said this, and looked at me, laughing. "Or maybe, after talking to Gary, you would."

"The language of smells?"

"You bet."

Cobby is athletic, tough, a strong condensed man who gives an impression of wary springiness even when he's sitting still. He wears a battered leather hat with red-heart beads and a broken arrowhead strung around the brim. Everywhere he goes he carries a double saddlebag made of his own home-tanned deerskin with a "survival kit" inside: a flint and steel and a knife. He traps all year; furs in winter, problem predators in summer. He traps what comes, but cats are his specialty. Cats are what he admires the most.

Once when we were talking, Cobby asked me:

"Don't you think trapping is cruel?"

"I don't know," I said.

"Well, I'll tell you," he said, "there are cruel trappers. There are cruel bankers, too. And cruel kings, aren't there? The fact is, there are cruel *men*."

There is something in the way he says this that brings to mind the fact that Cobby has been, among other things, a Baptist preacher.

"By the way," he continued, "a cruel trapper doesn't stay in business very long. It isn't to your advantage to be cruel. It's a sickness, and sick people don't have what it takes to be any good. I always think" — he paused — "I always think that the day I get hard about killing, the day I stop feeling something when I kill an animal, the day I stop respecting that animal, thanking it — you know, like the Indians did — that's the day I'll stop trapping."

"Cats are more visual than coyotes," Gary says now. "Well, you can see that with a house cat. Something arouses her curiosity and she'll watch, won't she? She'll look. Where the dog will sniff when there's something he wants to know."

Some trappers hang tin pie plates from a branch above their traps, or white rags, and Christmas tinsel is a favorite, cat-trapping season coming as it does right after the first of the year. Cobby uses feathers. Gary favors skunk pelts. Where a coyote would be repelled by catchy graphics, a cat is drawn in.

Gary uses scent lures, too, ones that he's made just for cats. Like the lures he makes for coyote and fox they're made to arouse curiosity, to appeal to hunger, desire for high social rank, territorial protectiveness, sexuality.

"An animal has four main areas of response," Gary told me the first day I was here. "And these are: curiosity, passion, social life, and food." He paused, stared at middle distance, considering. "In making lures, now," he continued, "you can branch off of these almost endlessly, and blend them together."

He stopped there. I could hear him stop himself. It wasn't that he didn't have more to say, but that he was in the position of a biochemist lecturing to a grade-school kid; he'd run so far beyond my understanding that he stopped out of kindness. But he'd caught me, I wanted more. This was it: smells, the vocabulary of the dark.

"It's like a language, isn't it? Smells," I said.

"That's it," he said, looking at me straight, suddenly, his eyes wide. "For them, it *is* their language."

Making lures is one of the things that Gary does well; so well, in fact, that he's gone commercial, selling his compounds to other trappers. Lure making is an exact science; its exactness lies in its components, its manufacture, and in the timing and placement of its use. It's a language with grammar, syntax, urgency, politesse.

He does his blending and aging of lures in the kitchen of the old ranch house. He uses an alchemist's profusion of ingredients: seal oil, glycerine, phenol, oil of spearmint, oil of catnip, valerian root, cinnamon oil, oil of bergamot, lanolin, skunk essence, pulverized animal glands, beaver castor, fishmeal, animal urine; other ingredients, too. The idea is to compound and age the lure and then "fix" the proper scent with sodium benzoate or another preservative. Some lures smell like ancient cheese, others like pine sap or musk; some smell truly terrible and all of them smell complex, layered and symphonic, even to the human nose. The biochemistry is similar to that used in human perfumes, and the goal is similar, to attract and titillate, but the audience is more sophisticated. Biologists say that a coyote's or a bobcat's nose is fifty-eight times as acute as ours. Where we smell only strength, the cat or coyote will smell specifics, and Gary's goal is to put specifics in. If smells are the idiom of the dark, Gary is a translator.

*   *   *

Now, in winter, the nights are long. The stars wheel, the wind rubs the country raw. The brush of fur on a branch leaves a message in the wood; a splash of urine in the gravel leaves a signature as personal as any name.

Gary's pockets are full of jars of lure and I can hear them clink as he walks. His recipes are secret. He improves on his recipes and experiments with these improvements all the time. When he reset a trap yesterday the last thing he did was to fish one of the jars from his pocket, bend a stem of grass, and scoop out a half-teaspoon of stuff that looked like peanut butter. It smelled like overripe cheese but with overtones — a spike of mint, or sweat. He poked the grass stem into the ground by the trap.

"What's *in* that stuff, Gary?" I asked.

He didn't answer.

Now we follow the cat tracks down toward the stream. They cut off the trail and disappear over the rock.

What took the cat off? I think of that: the shift of a shadow on stone, the glint of an eye in starlight, a whiff of crushed stems, a scuttle, a twitch, the bobbing scut of a whitetail jack or a cotton-tail moving off; I think of the cat sinking into his half squat, wagging his tail in curiosity and excitement, watching, ducking to sniff the stone, lifting his nose to taste the air, moving up over the rock as soft as a streak of shadow and crouching, then leaping, leaping again, leaping . . .

I shiver in spite of myself. It isn't the cold, though the air is as raw as grit in the eyes. It's the suddenness, the sudden leap and vanishment, of the cat.

The vanishment of the hare.

We come out on the creek bottom and Gary stops again.

"Know what that is?" He points.

"Weasel?" I say, looking at the twin-dot track in the snow.

"Mink," he says. "Come on. Hunts by the water. In and out. In and out. Mink."

He's already moving off.

"Beaver, too," I say, seeing what it was that the mink was

hunting in and out of. A series of dams is strung down the watercourse, dams no more than three feet wide, the gnawed wood white and fresh.

"Yep. A young fellow, just moved in." He laughs. "Beaver are good for mink."

There's a pause while I catch up. His walk is steady, his strides too long.

"You don't want to know about mink, too, do you?" he asks.

"I wish I had time," I say.

"Give me a year or two, and I'll teach you about mink." He grins. "You already know more than you think. All you need to do is get out there and trap."

"Me? Trap?"

"Sure," he says. "Why not? There's plenty of women who trap. Some of them are damned good, better than the men."

I think about this. I think he's trying to make up for shaming me about the gates.

I begin to wonder about my own country. I wonder if I could learn to read that. I wonder if I will ever see any country again without thinking: *slope, hollow, watercourse, rockpile, path, crossroads, scent post, circle downwind* . . . I wonder if I will see any country without thinking of maps and idioms in the dark; and I don't believe I ever will.

The valley widens as we walk and a hundred yards ahead I can see the roiled snow and a trap snapped closed and empty. When we come up to it Gary bends and pulls a twig from the metal jaws.

"That's all it was," he says. "That's all it took for him to pull out. Then, look what he did."

He points to the tracks in the dirt and snow. He bends, tracing the coyote's path with a finger, gauging speed — trot or gallop — gesturing to show direction. This is what he sees:

Last night one coyote came through here hunting alone. He was the patriarch of the family that claims this valley.

He came down the streambed in the dark, crossing back and

Coyote
Canis latrans

forth over the water and curving downwind of the brush and trees. He came guided by sure knowledge of his own home place, which is like our knowledge of our own house. He would have known any sound or smell that was out of place: freshly dug dirt, a shifted stone.

He came up from the river and then back toward the west, at an easy trot, patrolling the height of the draw along the bottom of the slope. Gary had set his traps along the bottom of the slope, knowing that when the wind was right the coyotes would come. The lure, compounded to arouse curiosity and territorial alarm, had been left in a natural hollow in the earth. The trap had been set in front of the hollow. The coyote came, slowly, in. He'd caught a whiff; he wanted to know: What was this? Who was this? Then he stepped on the hidden pan of the trap and the trap had him but the trap also had the stick, and the stick wedged the trap jaws open. The coyote ran the length of the chain and he was out.

"Then," Gary says, "he brought the whole family around to see what it was that bit him."

Here they came in. Here snow and earth are trampled flat. They came in, one by one, tense, nervous, bowing with haunches in the air, paws and nose forward, sniffing at the trap. Here the coyote dug, and dug again, showing: This is how it bit. From here. From underneath. I can see him in the moonlight digging and digging, running back to the others to see that they understood, then digging again.

We go on then to the next two traps. The pack has been to these, too. In front of them both the snow is trampled and the earth has been dug.

"A digger," Gary says. "That's bad. He'll dig up every trap in the place. Well." He tosses the twig in his hands, and looks grim. "We sure gave these coyotes an education."

We go back up the creekbed in silence. Under the saplings on the slope are the tracks of wild turkeys. On the open slope are the feather marks and crabbed pigeon-toed track of an eagle that has landed and caught a hare, leaving bits of fur and blood in the snow; at the end are the labored feather marks of *finis*.

We climb the ravine and get into the truck and drive on across the prairie. We go down through a draw filled with leafless aspens in which a log cabin is collapsed like a child's discarded toy. Above the horizon a lone hawk circles, like an apostrophe, giving shape to a high and hidden wind. The size of the sunlit land is large enough, but in some way that is impossible to describe it is not as large as the night, when the air has a columnar stillness as if stars held it vertical like strung molecular beads. The eye can lie. If it weren't for the tracks stitched through the scant snow it would be easy to believe that — except for the apostrophic hawk — this country is empty of life. The success of the hunter does not lie, and I know this: for every coyote that Gary captures — and there are hundreds every winter, caught from the same ground year after year and replenished every year like the flow of a river — for every one of these there is, in turn, a flow of what the coyote has fed on. This is continuous here, a wheel, and can be stilled only by poisoning or by paving over, though even then some spinning persists: there are coyotes even now in the city limits of Los Angeles, and larger, quieter, eastern coyotes in the Connecticut suburbs, in New Jersey, Delaware. They weren't always there. That they have predators, too, has not stopped coyotes from spreading their range farther and faster than any animal in North America except — some argue — for the clever, insatiable, night-loving raccoon.

We talk about these things as we go, and we talk about other things, too. Some of them are difficult things. I like Gary's long silences. The wheel of the prairie and the snow and the mountains and the darkness hovering over the rim of the world have the space in which to enter and sing in the mind. I know that Gary will out with it sooner or later; that he will pierce my question with a question of his own, with something more honed, precise, for me to batten on in turn. I get better, too, at opening and shutting gates. The truck stops and I wrench at the stubborn door latch and jump down and out; I have to hug the posts to my chest, pull the wire bales until they stripe my palm with red. Then we're through and the land tilts again and goes stony again and precipitous, and we come down to the Little Missouri along Dry Creek.

How do the coyotes and cats *see* this country? How do they know it themselves? The endless questions.

The creek beds wander and intersect the Little Missouri here in a thirty-six-square-mile area of prairie and eroded badlands. It's a communal cattle range in summer, Gary's trapping grounds in winter. It's a landscape of dark and light; black of trees and white of snow, grays of stones and branches, coppery duns of bluestem grass and scoria rubble.

Black and white and gray are the colors that the coyote sees, the only colors the bobcat or the fox sees, the only colors they can see. They have no retinal cone cells that subdivide the spectrum and sense an array of color; they have only rod cells, which respond to quanta of light. They see only light and dark, but they see better than we do in dim light; not only because they have more rods but because they have a kind of mirror — the tapetum — behind the retina. This reflects all incoming light back out again so that what light there is is sensed twice, coming and going. Other night animals have these, too — deer, coons, mice, cats. Shining a light at night, into the brush hollows and the aspen thickets, one can see those twin mirrors shining back — greenish, yellowish, red, rose — the colors, the spacing of the mirrors, the bobbing of their motion, giving form to otherwise invisible life.

These coyotes and cats see the country, even at night, not much differently from the way I do now. At night, with the backdrop of the snow, even I can see it well enough. If I think of my own night vision as a black-and-white TV with poor reception, then theirs is like a black-and-white TV with good reception.

Their hearing is skewed toward the high-pitched sounds of small mammals, sounds that I hear poorly if at all. Their noses give them a four-dimensional map of the land, and clues to the sexuality, moods, travels, even the intent of the other inhabitants. According to Bill Austin, who has learned to decode their songs, the coyotes' calls to one another say more than anyone would think.

I remember Bill Austin telling me — in the pale gritty interior of his pickup, the pickup parked in the evening light, his cup of coffee parked in the angle of thigh and seat back — what it is the coyotes say to one another in the dark:

"Often all you hear is this. We're talking about a male and female now that have pups. Long about midnight you hear that old dog go —" Bill makes a single howl, a round burst of sound with a high trailing edge. "And what he's saying is: 'Where in the world did *you* go?' Then you hear —" Bill makes a high, screechy, shaky yammer, "and that's her, see. Lettin' him know."

Bill grins, a lopsided grin that goes with his cowboy lankiness. He's been a rodeo rider, too, as Gary has. He's a pro trapper, though for reasons of his own he's no longer full-time in predator control, working between the anvil and hammer of government agency and sheep farmer.

"They told me my job was to kill coyotes," he says. "I told them my job was to kill *problem* coyotes. They couldn't see the difference." He shakes his head.

"Well, to go on with the coyotes talk. The night passes, see," he continues, "and about an hour or so later, and from somewhere else in the old dog's range, he's been moving all this time, you hear —" Bill makes the male's querulous howl again, louder, and it sounds wild, lone, dark, insistent. "That's all you hear. What he's saying is: 'Why don't you get yourself out of the brush and get up here and keep up with the outfit, we've got kids to feed, you don't have time to be playing around down there doing whatever it is you're doing.'"

He's serious. He enjoys this. This is no bloodless scientese; this is alive.

"Then, maybe another night you'll hear something else." He

considers, takes another pull of coffee. "Like, let's say it's come to be near dawn now, and she's got back to their den but he's not back and the kids are hungry, what the hell's he up to? Playing pool or something? So she'll get up with —" He breaks into a wild caterwauling, a tomcatty nasal quaver that is clearly frustration, urgency, even insult.

We laugh again.

"How about their territories? Do they talk about that?" I ask.

"Sure," he says, "when they want to. A full moon night there's a lot of talking back and forth. They'll say everything, it's a kind of partying, just like we do."

"What do they say?"

"Well . . ." He considers. "Let's say there's a bunch of coyotes here and they're a family. When that moon comes up the old dog sets down on his haunches and goes to howling and lets the rest of them know that he is the *king* of his territory." Bill howls again, but this is different; long, wild, complex. "That's the domain call." Bill smokes, sips coffee again. "When he starts that up and his wife is with him then she'll chime in and put in her two cents."

The wifely howls he makes are higher in pitch, full of musical verve.

"So then," Bill continues, "over here suppose there's a lone yearling. He might call back to them just to say: 'Yeah, I'm over here. So what?' It's a response call."

He gives me more howls, silences, explanations. When I ask him questions he pauses before answering; before he howls there's a slow breathing emptiness as if he were shifting himself out into a foreign tongue. It reminds me, later, when I think of it, of my own attempts to translate the poetry of García Lorca. There's the same breathing pause before the music, which is, itself, untranslatable. Plunked into English, some of it sounds ridiculous; there's no way 'round it, it won't wash. Read aloud in Spanish, that same poem will bring tears to your eyes. The grief of a translator is what I recognize in Bill.

Bill does his coyote research alone. He drives his truck in over ranch roads and camps there in tents and in the open, listening for coyotes after dark or before light; calling to them so he can listen to them calling back, and calling them in so that he can watch them, sometimes by the hour. Besides his binoculars he has as his only tools two whistles he's made himself, small plastic items bound by rubber bands, like tubular candy-colored panpipes. They bear a superficial resemblance to those penny-a-dozen prizes given out to children at school fairs. They also bear, not superficially at all, a resemblance to the laryngeal equipment of coyotes. He holds them out to me in the palm of one brown seamed hand as a juggler might hold the three oranges, or the six walnuts, that do not juggle themselves. Without a lot of listening to coyotes and deft practice at imitation these whistles can't sing; but by varying the strength of blown breath, muffling with fingers or palm, applying tongue-tip pressure to palate and teeth, Bill can, with these, call questions and whistle rippling replies; he can, with these, divine the whereabouts and sex and intentions of almost every coyote within earshot of wherever it is he has camped himself.

Bill makes most of his living now by teaching other people what he's learned.

"Each of those coyotes is an individual," he says. "Most people I try to teach want you to give them a recipe like you're going to bake a cake. You do this, this, and this-and-that this way and so on. You can't do that when you're dealing with individuals." He puts the whistles back in his pocket, drains the coffee cup, looks out the window at the prairie that rises beyond the road in darkening dun slopes. "You don't talk to two people exactly the same. You try to *relate* to people. You try to relate to the individual involved. And when you try to relate to a coyote, you have to relate in a meaningful way that will make him respond. Does that sound like women and men? I mean, it's the same doggone relationship! You have to relate to *that* coyote."

He sighs, smokes; the prairie fills with shadows.

"I detest statistics," he says. "They don't mean anything as far as I'm concerned. 'A coyote in this mood and this relationship and this situation will probably do this . . .'" He laughs. "What the hell does that tell you? You could say the same for women and men, couldn't you? But how is that going to help you when there's a particular man you want to get to pay attention?" We laugh, we fill our coffee cups from his thermos, and Bill goes on: "But there's some things you can say. A moonlit night is better for calling coyotes than any other night. A male coyote without a mate will howl very little. A female coyote without a mate will not howl at all. A yearling female without a mate is the hardest coyote to call in. She's not aggressive enough. The easiest coyote to talk to is the biggest, smartest dog in the whole country. Think of people: if you want a man, which is easiest to get? The timid one or the aggressive one? Well, now."

He grins and shakes his head. Dusk creeps down the slopes, the grasses go rose, then dark.

"For the most part," he continues, "coyotes just need other coyotes to play with. They play tag and wrestle and trail around. They need company. Company is what they're after. Their language is like ours. Each situation they need a different language, just like we do. Their language isn't as complicated. But it's just as expressive."

It's quiet here, now, along Dry Creek. I can see a long way up and down this badlands valley and there is nothing but stasis, the huge calm of land.

There are cottonwoods by the river. They look blasted, burnt, crushed, in the manner of plains cottonwoods. Without their greenleaf clothing they are all black skeletal struggle against wind, drought, bug-chomp and snow weight, against time. They hold, somehow, up. Now, under one cottonwood, I see the tracks. Here last night a coyote came to stand in the shadow of one leaning and blasted bole. Here above the water where the current chutters over a stone and sent up moon-crinkled light, a deer came down to drink. The coyote hunkered. Here is the brush of her tail as she crouched.

The deer drank and left. The coyote stood, shook blown snow from her coat . . . see the spatters here, here.

Then she went off down the creekbank along the pale indent of the ice, trotting easily with tail and back level and gait so even that she seemed to float. Now, see, her gait changed, lengthened; her head was up, nose, eyes, ears focused together in a forward aim. The crackle of grass, the jerk of a shadow, a scent of urine or breath; she circled, here, getting the picture the way a radar antenna circles, completing a whole. She aimed and rushed in then with lips curled back and teeth bared and jaws open and close to the ground, she came in fast and low — haunches high and head low — with the speed and shape of a furred bullet. The meadow vole was snatched from his runway as if he were hit by a swinging bat, was shaken twice, carried off.

Here, here, back along the cottonwood tree, around and arrowing off over the solid ice, she carried the vole away with that same smoky even-gaited drift that is one with the shadows, the air.

I am breathless with my own pursuit. I stand up panting. Gary has disappeared.

"Gary!" I call.

"Here!" he calls back.

We go on together. He doesn't talk much in the time that we are there, checking the line by the creek, and he moves slowly. He is attentive, concentrating, every minute of that time. Once in a while he looks at me under the brim of his hat.

We get in the truck again and drive off on rough tracks, and then we walk the further creeks in the snow under their cottonwoods, and coyote tracks are as plentiful here as they have been everywhere else. The draws are trampled by wild turkey, the spattery trails of white-tail jacks and cottontails lace the thickets of buckbrush. Mink have been patrolling the water.

When I ask, Gary tells me that he has been trapping this piece of country every winter for more than twelve years. In the whole place he has set just twelve traps. I know that the pan which will spring each trap is the size of a silver dollar.

He says that he figures he will catch fifty percent of the coyotes in here by the end of the trapping season, as he has nearly every year before. Fifty percent of the coyotes in thirty-six square miles will step, exactly, on twelve silver dollars.

"How can they!" I say.

"They do," he says, soberly, giving me another under-hatbrim grin, a blue-eyed stare.

I know that the old resident coyotes and bobcats know just where to hunt and just how, but this skill seems incredible in a person — in any person. I've heard the stories; how foxes will follow grazing cattle to pick up mice and insects stirred up by hooves; how newborn deer are hunted, in June, pairs or packs of coyotes working together, one decoying the doe away from her fawn so that others can do the killing. A pack of coyotes will learn to chase a doe onto winter ice where her own skips and turns will, finally, bring her down. In winter the frozen borders of the Little Missouri are scattered with the carcasses of deer that coyotes have chased there. A fox will learn to hunt for nesting ducks on spring nights, killing the mother duck to feed his young but caching the eggs in holes he digs, tamping them singly into the earth, and remembering where they are; he digs them up again when the land is lean. A bobcat knows the jackrabbit's trails and will wait on a limb that angles just over the trail, a limb she's used before. She'll stay there poised to leap down. She knows where the wild turkeys gather, at dusk, under cover of brush, where she can jump in and trap them in the branches.

I think of the twelve buried silver-dollar-sized trap pans.

"But how do you know where to put them, in all this?" I ask.

"Well," Gary says, and there's a long pause. "Some part of it is instinct. You can study for years, you can trap the country for years. But you got to have the instinct." He smiles. "It's what the coyotes have," he says.

*　　*　　*

Our last stop for the day is behind a farmyard where there are haystacks and a pond. The water is clear of ice and the haystacks are full of signs of mice and voles. The barn is just over the fence. The mud by the pond is patted with fox and coyote tracks as much as it is by the hoof marks of cattle.

A road goes out from the farmyard, the ruts white with fine snow. They parallel the fence line and disappear on the horizon.

We walk up the road and over the hill and we both see the coyote at the same time. She's lying in a brown circle of earth, its radius the length of her trap chain. She's asleep when we see her, but she wakes and lifts her head, and she doesn't move as we come up to her but keeps that stare, solid and unwavering and yellow, as if she were bathing us in a kind of light. We stop ten feet away and Gary puts his bag down and unslings his gun and loads it, and the coyote puts her head down on her paws. She keeps her eyes open but I shut mine. The gunshot sounds as if a high thin wire had been plucked inside my skull. When I open my eyes I see that her head is tilted down and that her stare has closed.

We leave the coyote in the grass and go on up the hill. There are two more sets to check at the borders of fields and another one a half mile off in the brush of a draw. The pluck of the gunshot still sounds between my ears but the lesson goes on: looking down now I can see that there are three sets of tracks in the snow of the road. Gary stops as we reach a rise and points to one set going off and through a fence.

"The same animal came in half a mile back," he said, "and from the same direction. This tells me that it was a resident animal patrolling his borderline and going home."

"And these other two?"

"Dispersing pups, maybe." He stares down. "I'd say they were young coyotes, moving through. But the one we caught just now, she was a resident, a neighbor come to patrol the same boundary, to mark the other side of these scent posts." He pauses again. "I set the trap at the junction of two farm tracks, a crossroads where animals were most likely to pass. I used a lure made of coyote scent glands."

"A stranger's mark."

"Yes."

When we go to the truck, I carry the bag of equipment while he carries the coyote. She hangs down his back nearly to his knees. Her fur is the texture and color of the prairie grass: back line and tail tip are dark as winter buckbrush, head and legs are the russet of bluestem, her belly is the milk and cream of shadowed snow, her flanks are brindled like the flanks of hills.

Bobcats, prowling riverbeds and high rimrocks, wear the striped and spotted patterns of the stone.

The white-tail jacks are clumped colors of grays and whites, half into their winter pelage, like the snowed ground.

"The coyote is the same colors as the grass, Gary," I say.

He nods, and rewards me with a blue stare of complicity. I am sure of my own blindness now, here, even in the light; even more in the light than in the dark, when I have no choice but to leave my sight behind.

We get in the truck to go home and I wonder again what it is I've found. I find myself looking up as Gary does, as if my hat had a brim to look just under, letting loose a question, even at nothing in particular — the sky, a house we pass, a field — at nothing that answers. The only thing I really know is that my feet are cold and that we're going home.

"They talk about the balance of nature," Gary says, suddenly, in one of the odd statements that surface from his long mulls, a statement that has rattled my notions since as much as anything has. "Nature is never *in* balance. It can change with the weather, it can change in a minute."

We drive home with the prairie wrapping around us like a shadowed quilt, the clearing sky as blue as water in a bowl. The stars come out again, reversed and dizzying. The only thing true has been living itself; all my old book-found data sound hollow to me, rude, shrill; and now the night opens like a curtain ringing up. Gary stops the truck and nods, and I look where he's looking, and I see the coyote too, dark against the snow, whirling and then running upslope and whirling again.

The coyote stops and looks at us and dips his neck to smell the warm smells that rise along the ground. His eyes never leave us and his ears never leave their forward aim. At last he spins and goes through a fence as if it weren't there and he is over the rise and gone. He leaves a track of weightless grace, like a ballerina whose work in life is defying gravity, or in giving the illusion of defiance.

After dinner that night I say good-bye, which is difficult, because for some reason that I cannot explain I feel as if I am leaving home. I get into my rented car and drive east on a straight road.

I thought of many things on that drive and I have thought of them since, but these thoughts are as hard as if they were made of a strong but slippery metal, made to resist the form that words might give them.

Perhaps, I thought then, we carry our darkness like a cloak, hiding our true nature from ourselves. If that is true, then we can proceed only on a flat plain in which we may judge what is right and what is wrong, but only from the dictates of fashion and culture: we are like ships that skim the surface of the world in the broadest light, seeing there only what everybody can see, and this is not much, after all.

The night of the planet and the night of the soul are not so different. To the dark we've banished our ghouls and our crimes and our dreams and our terrors and our gods, because they are elusive, though they are everywhere. I think of Gary and Steve and Cobby and Bill; not one of them has been limited by what is easily seen in the world.

They know that what is obscure may still be examined; if not brought to light, then at least deciphered with a human patience.

Cobby, for instance: he can tell from the track of a bobcat whether the cat was hunting, or out for mischief, or just passing through; and whether it was male or female, young or old, resident or stranger. I know that this knowledge has been won by years of pursuit and failure, and by necessity, and by what he admits to be a passion. He brings this same patience to bear on his own nature; and perhaps because he is part Cherokee and part Iroquois and part European, he gives thanks to the gods of all his fathers and is proud.

The size of starred sky and dark rolling ground makes me think of things more durable than myself. It's no wonder we make sense of these stellar dots: they're like Rorschach blots, they allow us to project what is in us all the time. They're eternal huge forms that drive our lives and make the brightness of conscious intellect look small, temporal, scurrying, like campfire sparks.

I think of all these things while I am driving in the dark, alone and on a straight road. It's two hours before I come to enough to recognize that the constellation ahead of me is Orion, the hunter, bent over in the sky. Beside him is Sirius, the dog star, the brightest of all.

HAWAII   January

Because I have stirred a few grains of sand on the shore, am I in a position to know the depths of the ocean?

Life has unfathomable secrets. Human knowledge will be erased from the archives of the world before we possess the last word that the Gnat has to say to us.

— J. Henri Fabre, *The Life of the Spider,* 1916

By Chance

Another name, another telephone number, a few notations on a
notebook page: this is enough of a trajectory to make me leave in
the dark. I drink coffee, shower, dress, haul my bundles to the
airport, and climb aboard.

Before dawn we take to the air. The tiny prop plane holds five
of us, captain included. The control panel is no larger than the
dashboard of a car. We are on our way from Oahu to Lanai. It
feels as though we were motionless in a medium stranger than
air. I have been on Oahu for a week already, but this dawn flight
is my first glimpse of any whole, of the smallness of the islands
and the curved immensity of sea.

I came here to explore the reefs and the mountain rainforests,
and I've spent my nights in one or the other and sometimes in
both. My days have been filled with the bustling business of
interviewing people, burrowing through the university library,
organizing notes, sleeping a fitful and shallow sleep. Days are a
struggle with intellectual constructs and physical necessity; the
night is no struggle. At night I come fully alive. I've turned
some corner here, of my own. I'm awake in the dark, shucked of
the *maya* of surfaces and human fiddle. Now in flight this dawn
seems like an arrival: these could be my own wings. These are
the colors of my dreams.

It's like being at the core of a pearl. We're a dark thrumming
speck in a vast translucent sphere. Below us the water is glassy
calm and reflects opal light, streaked and spotted with coal-dust

shadow. We fly just below the bellies of the clouds. We see Kahoolawe, Molokai, Lanai, Maui, and far away Hawaii, the domes of both its volcanoes softly visible, roundly curved like the breasts of a woman lying naked on her back. The colors are shell-like, bubble-like, fragile; bright glass of the sea, white of clouds, pale blues, the softest yellows and pinks — *we're at the heart of a pearl,* I think. *I don't want to go down anywhere.* Then the sun rises slowly behind a veil of haze so that it comes up like a perfectly round fire-orange ball, and I can feel the earth moving forward as if our pearl were a rolling thing. The sun does not rise, the earth *turns.*

The islands below and around us are dark and virginal, mountainous, in their fringes of breaking sea. They are soft and silent and whole as if they had just now risen into being. They are as they must have been when the first creatures saw them; the first canoefuls of Polynesians, Captain Cook, birds coming down exhausted in a wind, a squirrelfish flowing in on a gyre of the North Pacific current, the mote of a spiderling parachuting out of a jet stream; a spiderling hanging from a hank of ragged silk, having given up hope long ago, or what passes for hope in the mind of a spider.

Jetsam. Life has arrived by chance. The darkness is filled with creatures that washed in or drifted down. This is miracle: even on a big, big map these islands are specks. From here, close enough to the sea to see mother whales surfacing with babies at their flanks, their twin breaths rising and drifting off like smoke, the islands are humps. If you happened to be drifting or blowing through the world, the chances are that you would not get here.

Once you are here, the Laws apply. The earth turns, the sun comes up and it goes down. Night opens again like a door.

Sea Change

I learned today that corals go to war. Some of them, like the mushroom and sausage corals, are better fighters than others. At night long sweeper tentacles emerge from their polyps and arc, wormlike, into the neighboring camps, carrying stinging cells that are as lethal as those of a Portuguese man-of-war.

After the first rush of adrenaline on hearing this news, I found that I was not surprised. Lions nowhere lie down with lambs and visible or not there is this kind of arm-wrestling going on at the borders of everything; but it is new to see the layered and quilted anatomy of the reef as a palimpsest of wars; of battles fought hard and in the dark.

Corals have the work of building calcareous monuments by degrees, so it makes good sense that their intricate constructions need to be built in peace and calm with boundaries defended. It is good to know that not all of them are as feisty as the mushroom ones, each of which has surrounded itself here with a hard-won solitude, a kind of moat. *Porites,* for instance. *Porites* corals grow in calm water and there are several species here: finger coral makes forests of knobby longish shapes in deep water, bracket coral puts out columns and fungi-like shelves, lobed coral builds gentle rolling mounds in the shallow reef. None of these are good fighters as fighters go. They compete among one another by a kind of immunological push, the borders of their colonies advancing and retreating in response to chemical seethe; to the rejection of what is simply recognized as not themselves. Here in the shallows the fine-grained mounds of the lobed coral flow, yellowy, interfingering everywhere, like heavy liquids poured down.

Now at midnight I float over these coral hillocks, my drift as effortless as if I were an albatross buoyed by southern trades. All I need to do is to adjust my balance in the medium that holds me up: sea rollers heaving in. A yellow diving light hangs from my neck and shoulders but weighs nothing here. I am smooth, storm-toss'd. This faith in night water must be some memory from fetal days; I could have been here forever.

The coral hillocks disappear beneath my belly, within inches of my knees, gone like a pastoral landscape. A cliff drops to a valley of rubble. I adjust the light to see down — down miles — or is it feet? I soar.

*   *   *

Only twenty minutes ago John and Aubrey and I came out over the empty beach and along the wall of black rock. John and Aubrey are entomologists at the university and they have lived here for years; they own wet suits and underwater gear, but they have never been in the sea at night. Neither have I.

I have been here at midday, twice now, to swim with the colorful fish of the reef and get used to the place, but coming in along the wall tonight it was too dark; too different. It could have been another planet.

We were all a little nervous when we came. I should be used to this now, but perhaps I will never be used to it enough: the sense of having come too far to go back. I thought about it there, standing on the edge, and I wanted to laugh to diffuse the imminence, but there was nothing to laugh at. We took off our clothes with our backs to one another while the rollers sighed against the rocks. The sea surface was a tossing silken dark like cloth over skin.

We checked our diving lights, squeezed and zipped ourselves into wet suits, left a white T-shirt on the rocks as a homing signal, and went over the pitted rocks and into the surge. Once we were in and facedown, that fetal instinct of flight took us and we were surprised at our own comfort; we came up happily to tell each other so, but this was a mistake. With faces out of water we couldn't see where we were or what else was there, our legs were suspended rootlike in the rolling dark, and I could hear the edge of panic in my voice and theirs. So we went down, quickly, again.

Now our lights shine yellow-green as if we were swimming in diluted milk. The landscape of the reef moves beneath us as if we were blown and dropped and carried easily by wind. We stay close together. We signal by tappings, gestures, touch. When I see the others in the light, they look pale and bloated, their heads and feet distorted by gear, their hair slick and fine and flowing like satin threads. When I bump against them, their skin feels coarse. I've lost track of what I am. This is all right with me.

The valley gapes beneath me and a single squirrelfish arrows by, red and glowing as a coal. The valley is full of rubble furred with greeniness, with patches of rose and red, with algal blobbiness, with pale humps. I think of the corals warring in the dark. War, sex: the essential bloodiness of being, as pervasive here as tide and salt.

Now the valley is passed and I'm over shallows again. The coral mounds look deceptively soft, like the surface of a bog, and they're riven by black traces as if someone had been scribing them with a black felt-tipped pen. These are the burrows of pistol shrimp. At night the shrimp are busy and the water is full of crackling. John thought it was his hair hitting his face mask and I thought it was the rattling of sand falling in against the rubble, but it's neither: it's the noise that the shrimp make, a sharp frying crackle, like popping corn. The water fills with it like static.

Pistol shrimp are little more than an inch long. They look something like silverfish and move like that, in sudden jumps. One of their claws is hideously huge and looks dropsical and wrong; that claw is their voice and their weapon. The thumb of the claw has a knob and the palm of the claw has a socket; pressed in and pulled out, this Poppit-bead cork-and-bottle device makes a very loud noise. They use the noise to defend their burrows, to stun prey, to fight one another. They're feisty. If you put two pistol shrimp in a glass jar they'll pop and jab at each other until one is dismembered and dying, though it's difficult to see that they've touched each other at all.

I've seen only the kind of pistol shrimp that lives under little stones at the water's edge, and this coral-burrowing species looks very much the same, so they tell me, but it practices a kind of agriculture. Algae grow at its burrow borders and the shrimp harvests this for food. Though I can't see a single one I know they're here; the crackling is the noise of their language, like the racket a city makes.

The pistol shrimp have no monopoly on noise. Aristotle recorded his observations on oceanic voices in his *Historia ani-*

malium, which was published in 350 B.C., and he writes there about the piping sounds and grunts of some species of fish, and their "inarticulate squeaks . . . like the cry of a cuckoo," but his observations were largely ignored until less than fifty years ago. Only in the 1940s did people begin official listening underwater, and then they were listening for motor sounds of enemy subs, mostly; only after false alarms became epidemic was it understood that what the nervous naval officers had been hearing was the noise of life. This sea noise was loudest at night and louder still during certain phases of the moon. The navy's underwater microphones hummed with syncopated voices in the dark, caroling and urgent, like choruses of peeper frogs in spring pools.

Nearly everything down here makes noise of some kind. They have voices or instruments to orient themselves, intimidate others, defend themselves or their territory, communicate with clan or school, coordinate the release of eggs and sperm into the proper tide. Barnacles snap their beaks on their shells to make a crackly clicking or play castanettish tunes on their opercular valves. Lobsters growl, ghost crabs hiss, fiddler crabs beat their breasts, black mussels twang their byssal threads, sea urchins snap their spines. Fish grind their pharyngeal teeth to make a rattling grunt or beat against their swim bladders with fins or muscles to make percussive sounds: chirps, grunts, croaks, burps, beeps, booms, bumps.

Sea mammals — whales, seals, dolphins — make noise, too, but somehow this is not so surprising as the snapping of barnacles is, or the harping of mussels, or the saltwater catfish who feels and tastes his way with his barbeled chin and who signals in the darkness with war-dance drummings, rhythmic as a tom-tom.

Aside from the crackling of the pistol shrimp I'm not sure what I do hear. How much of this is waves on rocks, physical commotion? My ears aren't attuned to water noise anyhow, so it comes in blurred, fogged by a foreign medium, but what is not here is silence. All these bumps and hoots and cracks, this pervasive tapping: some of this must be language.

Unfocused as my sea ears are, my eyes work fine, and our lights work fine. We fly over a landscape that is surprisingly clear, but what I notice at first is only that so many of the colorful daytime fish seem to be missing. Then I see that some of them are still around, hunkered in dips and crevices, singly, but their colors and behaviors are so much changed that the fish are almost impossible to recognize. I have to stare, circling, identifying one and then another by shape more than color: it's the colors that are gone.

The convict tang are easiest to see, though their yellow-black striping has gone gray-on-gray with a line of pale spots. Their daytime schools have scattered and the fish have settled into single bedding places. They hang with their heads tilted down and they hardly move as I pass within inches.

Here a manybar goatfish sleeps, rocking slightly, in the sand. Then I see one butterflyfish, a threadfin; it spins and whips into a hole as I drift by, but not before I've seen that its bright daytime graphics of white/orange/black have shifted and dulled. Moorish idols, with their tiny pointed faces and long dorsal pennants, their daytime yellows dulled to sleep gray, drift up out of hollows as I go over. They are as sluggish and shapeless as scraps of weed and I can catch them easily in my hand; once they are touched, their yellow brightens and they wake and shudder off, away and down. In one crevice deeper than most a triggerfish hides. He's big, over a foot long. From the top he's as narrow as two hands pressed together and his rainbow brightness has gone mottled and brown; the only clue to his existence is the rhythmic rippling of his fins. When Aubrey dives to look at him, he backs, slowly, like a ship moving from a pier.

The colors have shifted to camouflage, the behaviors have changed from hunting to hiding, or to drifty friable sleep. The daytime fish are as animate now as weeds or coral rock. They rest half visible in coral hollows and the sigh of the tide.

What I wonder at more than anything is the numbers of fish that are utterly gone. I have seen only that one threadfin of all the hordes of butterflyfish that drifted around my head like petals at noon. There is not one wrasse to be seen of the bright schools that swam, at noon, within inches of my hands and face. Where are the parrotfish, big as watermelons? The yellow tangs, bright as lemony neon? The herds of big black triggerfish? Gone.

Some of the fish have just gone — in. The reef is full of holes, caves, tunnels. The holes between coral lobes and fingers are called *pukas* in the Hawaiian language, and it's a convenient term that every underwater person here soon picks up. I've picked it up.

The black triggers have pukas they use for sleeping in, and they'll use the same night shelter for months. They'll go back again and again until the oval opening is blocked by storm rubble or a congregation of sea urchins. These pukas are often the result of the burrowing of the pistol shrimp. As the coral grows around the shrimps' scribings, the hollows expand to form deep grooves, just the shape of a triggerfish.

In daytime the triggers range over the reef and duck into any hollow when threatened, but their sleeping pukas are something else. Two full hours before sunset the black triggers begin to flock together over the reef, and chase, circle, interact. It's social time. They drum their pectoral fins against their sides, banging their swim bladders like drums to make an "aunk-aunk" noise. When they're angered, they flash white patterns in the black: an aggressive warning sign. They zip off after one another and nip at others' tails. No doubt all this says much about their loyalties and pecking orders and their sexual identities and so on; if so, then it's a language that has not yet been decoded by us. Anyway, the squabbling goes on and as it goes on distinct colonies move toward their usual bedding places and begin to inspect the pukas there. Sometimes they try out several before choosing one. Choice spots are defended against interlopers. They inspect their pukas carefully and then slide in tail first until they're out of sight. The pukas are like bunks: safe, tight, small.

Once settled in, they hoist the spiny "triggers" in their first dorsal fins, and a spine on their bellies as well, locking themselves in place so that nothing can pry them loose.

Nothing will try to, except an eel. Even then a triggerfish can nip the eel on its soft nose. A big eel, once educated by a nip that can crunch the spines from an urchin, won't bother anymore with hunting in triggerfish pukas.

Some bed down early and some go late. Bedtime is a matter of individual taste. Some settle in as much as three hours before the sun goes down; two hours later half the colony has sheltered and the fuss begins to wane. Half an hour afterward all the triggerfish are gone.

Meanwhile the convict tangs and Moorish idols and butterflyfish begin their color fades and their drifts into caves and hollows. The wrasse go mottled and seek their own sleeping holes, or bury themselves in the sand and lie over on their sides, hidden till after dawn. The sand-burying sort often use the same patch of sand to sleep in, night after night. While all this is happening here in the shallow reef, other and stranger movements are beginning: an hour and a half before sunset the parrotfish and other tangs and surgeonfish move out into deep water. Their movements, like the others', are cued by the waning light.

The yellow tangs gather and mill where the reef drops steeply away. Then they go off along the lip of the reef.

They follow paths. This is the wonderful thing.

The paths they follow are consistent, night after night. If you wait there at the right place, at the edge of the drop to deep water, you can see that their movement is as regular as the passage of city commuter traffic on subway or highway or train track: here they come. Thirty-four hundred yellow tangs will swim past your face mask in an hour and a half.

The yellow tangs aren't the only ones who have a "crepuscular parade," as Bill Walsh calls it. Bill is the one who discovered that this happens. He watched it happen over months and years, hanging there in his scuba gear over the edge of the reef.

The parrotfish do a parade, too, moving higher over the bottom. Goldring surgeonfish pass five meters below, on the drop-off slope. Butterflyfish, goatfish, other surgeonfish, and other tangs also pass, each at their own depth and on their own trajectory: on their own paths.

Winter is spawning time for the yellow tangs. The first quarter moon is the peak of winter spawning time. Now, here and there along the parade route, a male yellow tang will claim and defend a place of his own and will wait, shifting in the flow, circling, swaying. When a female passes him, he swims, quivering, in front of her, fins erect. He flutters his tail and swims up and down again, dancing over and around her. If she is receptive she will raise her fins and they will rush upward, together, shedding their eggs and milt.

Afterward the female will go on to her sleeping place. The male will wait in his claimed spot and dance for others. When the parade thins after sunset, he will go, too.

Eggs and milt meet in the water and the embryos grow and the larvae hatch and drift with the tides and offshore gyres of current. Months later these larval tangs, grown to tiny fish, will fall out of the currents and gather in colonies of finger corals. There in the deeper reef they will hide, feed, grow. When they grow to adulthood, they will join the parades and will move with them for the rest of their lives, on the daily commute to and from the shallows. Only the adult tangs leave at dawn, parade back at dusk. Where they move to at dusk makes utter sense; they go back to the colonies of finger coral where the currents once brought them and where they grew and matured; the familiar peaks and valleys of home.

This yellow tang parade begins an hour and a half before sunset and peaks at five minutes after sunset, day after day. At the height of it all their plain neon yellow begins to change; a white bar emerges on their flanks. This color shift is a kind of advertisement of aggressive motives. They do battle for sleeping places as triggerfish do. Once over their finger corals, they chase and dash, the white bar glowing. As they settle into pukas a black patch and bar appear in the yellow; a prelude of sleep, or signal of waning aggression, or camouflage; perhaps it's all of these. Half an hour after sunset the water has darkened to cobalt and the yellow tangs are gone.

Now in the shallows there are creatures that we have never seen before. Except in an aquarium, perhaps, except in books. As we drift and sway, our lights pick out one, then another. Everywhere are the red flashes of the squirrelfish, translucent ruby red. They barely care to swim away. Some are as small as goldfish in a tank, others as large as goldfish in a pool. All of them are a bright clear red with silvery highlights and large dark eyes; they are not shy of me but hang alert, watching. They hold their dorsal fins up like a row of spines. Some of them flash white patches of alarm on flank or tail-base when they first notice me, then these fade. In one place I see a bigeye, a redder and larger fish with a huge black eye and a mouth tilted comically upward; he zips under something so fast that it's only a glimpse, like the flash of a scarlet flower in blowing leaves.

There are cardinal fish, too, in pinks and blues and silvers, their fan of transparent fins picked out in white and black. Compared with the squirrels the cardinal fish are fragile and iridescent, the colors of a sunlit soap bubble or an oil slick or that silk which goes orchid or mauve when it's tilted to the light. They are not shy of us, either. The squirrels hunt the reef surface and the cardinals stick to the overhangs at the fringes of the sand, and in deeper water I see another bigeye (or I think I do; the deeper red and the rounder body), and I think it's a bigeye most of all because it scoots so quickly off.

There's more; more wherever I look. A spiny lobster appears over the rim of coral and jerks backward into his hole, leaving his long antennae out and waving them to taste and feel our presence and our motion. One cowrie glides under a rock ledge, feeding. A porcupine fish glides by, as big as a cookpot, his eyes milky in our light; a nocturnal hunter of crabs and mollusks. Here a zebra eel lies twined in the coral. He is busy poking his long snout into one puka after another, intent on his hunt. A black sea cucumber, thick as a pipe and covered with sand grains, navigates the sea floor. A solitary trumpet fish, as long as my arm and as slender as a bone, glides facedown over the sand; he is almost colorless and nothing of him moves except a minuscule frill of hind fins, which vibrates like a motor. His long jaws are stretched into a pouting tube and his big eyes search the sea floor; if a fish comes in reach he will open his jaws and suck it in. His stalking looks like the drifting of a stick.

In daytime the squirrelfish and cardinal fish and bigeyes rest in schools, in caves and under ledges. Dusk and dawn are the changings of the guard; at dusk the cardinals are the first night fish to arrive in the reef. They are hunting the shallows by fifteen minutes after sunset. Twenty minutes later the squirrelfish boil up from their caves and move to the reef and divide and conquer. All these fish take predictable paths back and forth from bedding place to feeding ground, just as the tangs do. Nocturnal fish are more predatory and more primitive in form than diurnal ones, so say the ichthyologists, and their colors are different from the poster-paint colors of butterflyfish or tangs: they live in washes of brown, silver, red. A lot of red. Red is the first color of the spectrum to be lost to the eye when light is dim. In dark night water red fish or crabs are — black.

There's more, more wherever I look: the rocklike body of a scorpion fish, barely visible in the sand, with a grinning mouthful of needle teeth and too-large eyes peering balefully upward; a brotulid curving over the bottom like a barbeled gray eel; the silver flash of a large jack, sickle-tailed like a tuna, its body shining like a mirror. It arrows over the reef, looks once at me, then turns away.

Then we're back near the shore and the surge steepens and the waves want to bang me against the stone. I pull off my gear in a single motion and breathe the air; the wind is cold on my face and scalp and there are stars. I put out a hand and hold on to the pocked black wall.

A volcanic wall. Suddenly I think of this; that what we have

Hawaiian Squirrelfish
Sargocentron zantherythrum

been swimming in all this time is drowned crater. The curve of land against the night sky is the curve of an amphitheater. The Pacific glitters out there where the crater's lip tilts off, and the waves draw me out in a sudden sweep. I claw at the stone, but the rock is slick and I'm swept away. I'm carried and bumped hard and I remember someone telling me of the night that she saw white-tipped sharks come in here and play, whirling with each other like big dogs, here where the water deepens.

Then John and Aubrey have surfaced, too, and we're together and we've got the wall with our fingers and then knees and then we're out, upright, on dry and broken rock.

The sea is strange again and empty below us. It's a moving surface like a curtain that lets nothing show, that goes out wide to the horizon. It was cold in there; we're cold now. We shiver on the shelf of stone and hurry with the towels and with dry clothes. Aubrey opens a beer and we pass it around, toweling our hair, and the beer tastes thin after the salt of the bay. We look at each other with bright eyes and smiles as if we had shared a victory — teeth show white in the dark and our eyes glitter — but we're still in the undersea habit of saying not much, of getting by with a touch on the arm, a gesture. We were there among creatures wholly strange, all bones and silks and flash, sparks of inextinguishable fire; lives still molten there in the crater's heart.

Armored

At night in and by the water we can see them where they live: big Grapsid and Xanthid crabs, hermit crabs and ghost crabs and sponge crabs and swimming crabs. Shrimp: Sarons and Callionassids, harlequin shrimp, candycane shrimp, cleaner shrimp, dozens of shrimp. Lobsters: moronic-looking slipper lobsters, black spiny lobsters specked with white and carmine, mole lobsters of sea-cave darkness, Hawaiian lobsters bristling with sensory hairs. All are inhabitants of burrow or puka-city or hollow;

all are night hunters or scavengers, equipped — in defiance of gravity — with eight to ten clawed or pitoned feet on which they potter like moon-vehicles, possessed of a faery whiskiness, an unrobotical grace. Their stalked black eyes tip and swivel; some have long arcs of antennae that wag, tap, sensitive to touch and smell and to disturbed and disturbing currents. The forked antennules between their eyes jerk and flutter like semaphore arms, waving the bushlike ends that are furred with esthetasc hairs. These are the sensory hairs with which they taste the water, like coyotes scenting the wind.

At ten in the evening we crouch under coconut palms with our lights shining on the sand. The performers are shy and we try not to shine the lights on them. Instead we spotlight a swatch of pale beach with a crushed paper cup, the lid of a suntan lotion bottle, the prints of tourist feet, the flattened patch where a damp towel was laid down and where the sand still wears the dual-scalloped imprint of a sunbather's rear.

We catch winkings of eyeshine and hear muted clackings and soft scufflings and I stifle a laugh; not that the daytime show wasn't funny enough: the parade of scented grease and hundred-dollar sunglasses, the scorched skins and overheated egos, the whole spectacle of beachy vanity in which I have been, from time to time, a willing performer. No no. It's what went on, underneath; what is going on now that the people have gone.

We crouch, watching, bare feet in the wet grass.

"There's one . . . ! No . . ."

"There?" John says.

"Yes!" Aubrey whispers.

"Yes! Look at him *go!*"

Ghost crabs are the color of the sand. Most of the time they run too fast to catch, scooting sideways or backward in sudden zigs with their claws folding in across their chests; it reminds me of the elbows-up sideways float of a dancer. They wrench into their burrows like water down a drain. Their burrows are the size of eggs, more or less. Join thumb and index finger at the tips and you have it — a lozenge the size of carapace and tucked legs. Already the sand is pocked with burrow mouths and with the fresh stitchings of scuttling feet. If you walk just anyhow along the beach at night, shining the light, all you notice is that here and there the sand itself seems to creep; it could be the shift of shadow except for the ruby flash of live eyeshine. Here and there is a heap of loose sand like something left by a child.

All day long the ghost crabs have been deep in the cool and dark of their burrows and people have been walking over them, lying over them, reading murder mysteries, and rubbing each other's backs with coconut oil over them.

"They've been here all the *time!*" John says.

"Damn!" Aubrey laughs.

The male crabs throw up their castles at dusk and guard them all night, using them to attract females. The bigger the castle the greater the attraction, and one of these heaps is nearly a foot high; an eminence. The crabs do their feeding at night and are opportunists by trade, picking through seaweed and other detritus and catching sandfleas and flies. They like to scavenge meat: stranded fish or shellfish, a crust of tuna sandwich, a hot-dog. Like beach tourism itself they're almost everywhere that is sandy and isn't cold.

Along the cliffs after dark we find big Grapsid crabs that are the color of the rocks they're on: black as volcanic stone and specked with constellations of camouflaging creams.

These Grapsids aren't shy. When we walk the path just above the breaking waves, we nearly step on several, and they're big; some of the females have carapaces as broad as cantaloupes. Our flashlights don't bother them much, they cock their eyes our way and do the jerky scuttle-then-freeze that makes them so easy to miss. They feed in the splash zone and have spoonlike claws in front to scrape the rocks for algae. Their legs are tipped with pitons for clinging on vertical rock, for holding on against the suck of the sea.

The Hawaiians call these Grapsids *a'ama,* which means "to offer," and long years ago they were used as offerings to the old gods. The Hawaiian fishermen caught them with a Y-shaped stick that had a stripped coconut midrib lashed across the open top of the Y. Creeping up on the crab (this isn't hard to do), they placed the midrib just by the socket of the crab's eye. Then the crab was disturbed by a shove or a shout.

When he's going about his business, a crab's eyes are up on stalks, but a disturbed crab will clamp his eyes down into their sockets in order to protect them as he scuttles and shoots toward a crevice. The eye would clamp on the waiting midrib; the crab would be caught by its own reaction. The more he struggled the tighter he'd be clamped since the clamping of eyes is an alarm response, out of conscious control, like the raising of a mammal's hackles. The clamped crab could be popped in a bag, banged with a stone, offered up.

After watching them for hours it comes to me at last why it is that crabs make me laugh so much: it's all in the arrangement of their eyes.

If they see me they look at me with a fixed stare, no matter what else they're doing or where they're off to.

It's like watching someone cooking dinner or getting dressed while goggling wide-eyed at the ceiling; a pratfall seems always in the making. I wait for them to trip, topple, walk off a precipice. While they're looking at John or Aubrey, the big black Grapsids often run over my feet, and even pause there, gripping my ankle, tempting me to jump and yell. All this upward looking makes sense to crabs; crabs are always belly-to-rock (or sand) and their predators have to come from above, shadow shapes moving in against the lighter sky. With eyes that are flexible and stalked they keep watch on the world, like lookouts in turrets, and once they've sighted danger they keep it fixed in their sights and navigate to safety by touch, taste, memory, smell. They scoot and then stop, hoping you've lost them, that they've confused you with scoot-and-stop motion and that camouflage has worked. The space between my feet seems crevice enough, to them.

Underwater, there are more. It's daunting still; the dark tossing of swells and the cool of the midnight wind, the surface eddies that show by their swirl the shallow outcrops over which one cannot swim and against which one can be bashed. The water is cold, cold on my toes, cool sand conforms to the hollows of my feet. There are lumps of coral underfoot like files. Now we're in to our thighs.

We follow Marj Awai, a scientist from the Waikiki aquarium and a longtime veteran of night dives. Before she was officially a scientist, when she was a Hawaiian girl who loved the sea and spent most of her time in it, she used to collect reef fish on a commercial basis. Now a net is tied to her waist and a metal rod is lashed to her flashlight, simple enough gear for poking things out of holes and catching them up. Most reef fish are torpid in their pukas now and are easier to catch. Most of the shrimp and lobsters and crabs, she's quick to point out, don't emerge from their burrows at all except in the dark.

Even a full moon will keep them in, she says. The dark of the moon is the best night-hunting time.

This is that time, and this reef off Honolulu is a good hunting place. Behind us the aquarium is closed for the night. When I look up and back from looking down, I see only the line of Waikiki high rises lit like crates of fireflies. Just in front of us I see the long white flail of a conger eel disappearing fast into the dark bulk of reef out there. Four feet long if he's an inch. I pull on my flippers and rinse my mask and spit into it and rinse it again and then I'm facedown and all night strangeness goes, shifts into the strangeness of the sea.

The water fills with a familiar congregation: squirrelfish of five species that I can't separate yet, only noticing that some are more silver in the beam of my light, others more red, others subtly striped. In scoops in the reef groups of daytime fish hang, mottled and darkened: milletseed butterflies, Moorish idols, all floating like tethered kites. A speckled moray peers warily up at us. He's curled in the reef like a cat. Marj says he's always here at night, this one eel, patrolling his piece of reef.

The eels — conger, morays — do eat some fish, but most of all they feed on shrimps, crabs. Their long snouts are made for puka-poking. Their strong jaws are made for the crunch.

Inches from my belly a red swimming crab comes up and sideways in a smooth shiver, the hot color of boiled crab, a pinkish red, like sherbet. Its legs are flattened into paddles so that it seems to float, dippily, like a butterfly, almost colliding with me as it flows away.

Then there's a slipper lobster, mossy and spotted as a coral rock and curved like a rock and known by many names: Spanish lobster, shovelnose lobster, *ula-papapa* (flat lobster). Though it's the length of my forearm, nothing about it looks lobsterlike in the least except for the meaty tail. It has flaps instead of front claws, it has no arc of antennae, it doesn't have stalked eyes, either; its eyes are small and fixed on the sides of its head, like a whale's.

With antennae modified into flattish bits around its face it looks almost the same — curved — fore and aft. The ula-papapas rely so much on their camouflage that once you do see them they stay put and are easy to watch; except for the wagging of antennules they look hardly alive. I've been told that once alarmed they'll contract that vast hind muscle, and instead of scooting backward as a spiny lobster will, they'll use their forward flaps as ailerons and will somersault up and over your head.

This shallow-reef slipper lobster feeds on sea anemones, small shellfish, marine worms, and small crustaceans, whatever he can find. The deep-sea slipper lobster eats shellfish almost exclusively, and his legs are like wedge-shaped talons specialized for the shellfish trade. He can taste the effluent of an open oyster with his antennules and will jump and wedge his first set of legs in the shell to keep it open, then pull the shell wider with his second set of legs while he severs the oyster's adductor muscle with the third set, relaxing the animal's shell-hold just as one does with a clam knife.

The slipper lobsters are secretive, hiders and sneakers, nocturnal almost always. They're good, very good, to eat; I've eaten them myself in a Spanish form of bisque on which one can spend an entire afternoon of gradual guzzling. Sharks and rays like to eat them too, and octopi; even the big silvery jacks know a good thing when they see it. If they see it.

Most crustaceans are difficult to see. The Saron shrimp, for instance, are wonderfully shy. When I peer in their pukas, all I usually see is eyeshine and candycane bristle. Now in the dark of the moon they hang half in and half out of crevices in the coral rubble. They are as nocturnal, Marj says, as anything gets. On moonlit nights they stay inside. They're colorful and large; their eyeshine deep red, their legs striped boldly red and white, their curved bodies as blue-green as polished marble. The males have long striped pincer-tipped front legs and they fight each other by joining and pushing, like male deer locking horns.

My favorite crab of all is the hermit we find there on the reef off Waikiki. He's the size of a man's clenched fist and the color of raw meat; you can't miss him. He's living in a triton's trumpet shell, the attenuated shell-tip cocked rakishly sideways like the feather of a fancy hat. His heavy red claws clamp over his chest as we pick him up, his antennae stick out like bristling brows, and his eyes peer out, black as drops of ink, under the curving brim of the shell. The shell is plastered with reddish jellyish mounds, which prove to be sea anemones. They open slowly like the paper flowers you buy in Chinatown, the kind that unfold upward in a glass of water.

It turns out that this particular kind of anemone likes living with this particular kind of hermit crab, and that the feeling is mutual. The anemones provide protection and camouflage; the crab provides mobility and food; scraps of the crab's meals drift into the tentacles' reach. The trumpet shell gives the crab a mobile puka and the anemones a free seat. When the crab outgrows his old shell and finds a new one to move into, he will take his old anemones with him. When he taps one at its base with his claws, it will unseat itself, and the crab will pick it up and place it overhead like a floral decoration. The anemones will latch on there and off they'll go.

Later I found this out: that in the deep-sea hermit crab this business of workable partnership has gone farther still. A young deep-sea hermit will collect an anemone and will seat it on his shell, and as they grow together the anemone's leathery body makes a "shell" for the hermit to live in. The original shell gets buried in their mutual bodies like a bone. Unlike shallow-water or land hermits this deep-sea character never needs to househunt for a new home; the anemone is his home and he is its occupant.

Is the deep-sea hermit-anemone one organism or two? At what point, really, does relationship define a whole?

There are enough examples of this kind of tight-knit living to drive you crazy if you care to be so driven: hermits and anemones, farming ants and their fungi, ourselves and our mitochondria, corals and their zooxanthellae, termites and their gut flora, to name a very few. Evolution looks more than anything like a flamelike process; not only of endless subdivisionings but also of joinings up. The Linnaean system of pickling things in separate jars and applying separate names to the jar labels has its place, but has its limits. Some things won't submit to being cleanly divided with a terminological hacksaw. Some things are in the process of becoming, some things are already, irrevocably joined.

The possibilities are infinite, so it seems. Anyway I never tire of looking, especially here, especially since I've heard since the earliest years of my biological infatuations that shallow reefs are the most productive environments in the world. Warm water and sunlight make for a seethe of phytoplankton, and coralline algae, and the specialized algae that have joined up with corals: the zooxanthellae. More sunlight is captured by these micro-plants on an acre of reef than is fixed by an acre of rainforest or hybrid corn; to live in a reef means to live in a lapping stream of nutrient ore. Here as anywhere form and function define fit.

Fit: niche or puka, place and method in the seethe. This is city living. Here as anywhere one whole set of life-styles is tuned to the night: to the other half of the planet, to the rooms of the dark, which can be pried open here and now with nothing fancier than a flashlight.

It isn't just crustaceans and nocturnal fish and night-feeding corals, it's mollusks, too: trumpet shells and helmet shells and cowries, flowery nudibranchs, octopi, squid.

And other things. On some nights more than others the water is milky, you can see it when you dip your light; a mistiness. One night I focused on the mistiness and saw that it was moving. I saw motes: a bobbing mote, a circling mote, a squiggling mote . . . plankton. Arrowworms. Shrimplets that spend their whole lives tiny and afloat. Sea urchin eggs. Larval crabs, larval barnacles, larval mussels and snails, larval fish. There are also things that are not larval: protozoans, foraminiferans, siphonophores, and comb jellies, all of which look under the microscope like light-filled crystals more than animals.

The micro-plants of phytoplankton are the pastures of the sea, and these crystalline creatures are the grazers here. Tiny planktonic crustaceans, of which crabs and lobsters and shrimp are the adults or the big cousins, are like insects on land: the most diverse and most numerous of animals. During the day most of these zooplankters are mossily attached to the reef surface. At sunset, they ascend. If it is a moonless night they'll go down for a nap and then rise over the reef again early in the morning to feed again. They'll settle downward when dawn breaks.

By the time the sun rises the water is as clear as glass. At midnight I can watch these most miniature of armored things dancing in my light beam, a swarm like minuscule moths; drawn, spiraling, tossing.

Even these make me laugh.

*   *   *

I have my favorites now. First the hermit, then these. When I walk or swim over the shallow reef in daylight I sometimes see their holes. The holes are as empty and as perfectly round as the ends of pipes.

After dark two red claws protrude from these holes, spread and still like waiting hands. They're the size and shape of crayfish claws. Long antennae curve out between them, stiff and still like bristling hairs.

These are Callionassid shrimp. They make their living by waiting at the mouths of their burrows, flexed like the jaws of a trap. Anything drifting in, anything tasted by or touching the antennae or the sensitive whiskery hairs on the claws themselves, is grabbed and pulled down.

On nights when the tide is low we walk over the shallow reef with the water at our ankles and knees. We keep our lights in the water and hold our face masks in our hands like glass-bottomed buckets. John wears a pair of old boots for reef walking. One evening he pulls a bootlace out and dangles it between a pair of Callionassid claws. The trap springs so fast it makes us gasp — the shrimp and the bootlace with it jerk backward out of sight and the lace begins to reel away down with astonishing speed.

John hangs on. In less than a minute all that is left in his hands is the tip of the lace. When he pulls it out again, tugging to get it free, we measure it. The lace is twenty-six inches long. The shrimp sulks down there for fifteen minutes before it reappears, then it spreads its crimson claws gradually, opening in the dark like temptation, like a wicked flower.

Armed

The first time that I saw an octopus alive and close was the night that I went fishing with my father when I was nine. We were living in Italy then in a shabby hotel where there was nothing for me to drink but bottled water. I remember that. The bottled water tasted prickly and saline and I hated it then and still do.

I don't know how my father managed this expedition. He didn't speak any Italian and the fishermen who took us out with them spoke nothing else. One evening we put our old sweaters on and went to the beach and two men were waiting for us on the edge of the sand in an open boat; we nodded and smiled. They'd put a cloth over the middle seat in our honor.

I remember the stars; thick sprays of stars, and starry light dancing in the black water.

Once we were in the boat, the men rowed out into the sea and lit a hissing lantern and hung it over the bows. One man stood up with a trident in his hand while the other one rowed.

After a while I went to sleep. It was comfortable there, rocking and warm, with the hiss of the light and the slap of waves. Suddenly there was a lot of shouting and I woke up and there was the octopus: the octopus was as big as a beachball. It writhed on the trident like a glossy tangled pudding. The fisherman yanked it from the trident and plunged his hand into it and then his face.

"He's biting it between the eyes," my father whispered. "He's turning it inside out."

I remember tentacles going up the man's arm and over his ears. When he'd finished doing whatever he did he pulled the tentacles off himself and they came away from his skin with a noise like a hundred zippers.

Then he lifted a section of floorboards and threw the octopus in there, pushing the last whorls of tentacle from his hands as though he were scraping off suds.

The fishermen caught other things, too. They pulled abalones from the rocks and scooped a basketful of black sea urchins and they speared some fish, which slapped around awhile under our feet.

Hours later we went back toward the lights of the town. We were eating an abalone out of its own shell. One of the fishermen had carved it up with a knife and had sliced a lime and squeezed it in. Later he gave me the shell. I still have the shell, in which I can still see the cuts of the knife, faintly, like chalk marks in the opalescence. We were passing the abalone around and drinking wine and nodding happily to each other, when I noticed a trickle of tentacle creeping out from the floorboards and up the side of the boat. Then I saw more tentacles, splayed out of the jailing boards like rootlets. They were very long and thin and dark red. As I watched them they changed. They blanched pale, went a mottled dark, then flushed to red again. I crossed my legs up on the seat and said nothing. I drank some wine, which I thought was almost as nasty as bottled water, but which I had already learned to drink in self-defense.

Now here on the Oahu reefs I want to see an octopus very much. Why is it that childhood experiences are so graven? Somehow they are.

The octopi here are small. There are two kinds: a dark daytime species and a spotted nighttime one. They both live in the reef where they claim a home hole and adjacent hunting grounds as their own. They grow to five pounds at the most and live little more than a year: the males die after they mate, which doesn't seem fair; the females survive to brood their festoons of eggs, defending them against all comers, blowing water over the grapey clusters to aerate them, stroking the eggs with their arms to keep them clean. As soon as the eggs hatch the female dies, too.

Other bigger octopi can live for years. Most octopi are nocturnal and some live deep in the sea where it is always, more or less, dark. All female octopi put in time on egg care, and it occurs to me that this parenting behavior isn't the usual thing with mollusks. Bivalves spew their eggs and milt into the sea and let the sea take care of it. Most snails are hermaphroditic, one firing chalky darts deep into another's anatomy to signal a willingness for mutual exchange of sperm; afterward they both lay their eggs somewhere and go about their business.

An octopus is wonderfully strange. There is, for instance, the business of octopus eyes. Their eyes are almost identical in form to ours. There is no question of relationship: any common ancestor we share was extinct millions of years before eyes were the done thing, or before child care became something worth investing in. These things have developed, separately, because they separately work.

Then there is the fact that an octopus is an intelligent animal, intelligent enough to be fond, cunning, sneaky, even vengeful. There was the problem of the disappearing rock crabs at the Scripps Aquarium, for example. This is a good story.

A few years back the rock crabs at Scripps were disappearing at a steady rate, one every night or two. Almost every morning the aquarium workers would arrive to find another one gone. Gone: no shell, nothing. Finally, one person took a folding chair and waited in the dark aquarium after hours. The room was curved and the line of tanks faced inward. The rock crabs had one tank in the middle of the curved wall and an octopus lived in another tank twenty feet away, at the edge. Several large displays of fish lived in between.

Two hours after closing time, when it was good and dark in there, the octopus began to move. He crept, slowly, out of his tank and into the next, then into the next and the next. The tanks had no lids on them and were all open to the work space at the back of the aquarium, and the brief emergence into air while he clambered from one tank to another was no problem for the octopus. At last he was in with the rock crabs. He caught one crab with a snatch and flail and quick enveloping and then proceeded to take it home with him, tank to tank, halfway around the room. Once he was home he ate the crab at leisure, picking it apart with his beak and stuffing the bits of shell out of sight under a rock.

The problem of the rock crabs was solved by putting a rim of Astroturf around the octopus's tank. Octopi can crawl into and out of nearly anything and can squeeze their elastic selves through almost any hole, but for some reason they do not like crossing Astroturf. Once jailed, the octopus in question sulked for days, hiding himself among the rocks or plastering a mash of tentacle against the tank face, turning his hind end to the world at large.

Some octopi have learned (on their own, too, since parenting goes no further than hatching time) to use tools. One aquarium specimen used to sneak up on open mussels and pop a stone into their shells, wedging them open so that they could be easily devoured.

Marj Awai told me these stories. She was the one who discovered the crab-larcenist at Scripps. She likes octopi and treats them as friends, which they become. She enjoys telling stories about them, and her favorite octopus story is one about Bob.

Bob was a Hawaiian octopus that she captured herself. He was called Bob because he would bob up and down whenever any of the aquarium workers came near his tank. He didn't bob for the general public. He knew which side his bread was buttered on. The problem was that one of the young aquarium workers used to tease him. He would hold a crab out so that Bob could see it and then would snatch it away, laughing at the octopus's antics. He did this several times; Bob thrashed and squirmed. When he lifted the screened lid of the tank to put the crab in, Bob squirted him in the face.

One evening we go out onto the Ala Moana reef with Arnold Suzumoto, an ichthyologist from the Bishop Museum. When the tide is low, the flat-topped reef is like a wide pitted plain and one can walk there easily enough, though Arnold is always checking the landmarks around us, the lights of Honolulu and Waikiki, to see where we are. It's easy enough in the dark to lose the way, to walk off an edge.

We see a half dozen varieties of nudibranchs; one as lacy and beige as fancy underwear, another as round and orange as a half apricot. The cardinal fish and squirrelfish are out, too, singly, slivers of glistening predatory red. Red swimming crabs flow and scoot, rock hollows twinkle with the eyes of the Sarons, Callionassid shrimp splay their pink claws. Clusters of oysters have opened their mouths to the low rich tide.

This is what the octopi eat: crustaceans, bivalves, mollusks. One of their favorite foods is cowries. In the old days the native fishermen here lured them from their holes at night by dangling cowrie shells on strings. An octopus's reaction to a cowrie shell is as enthusiastic as Bob's reaction to the sight of the crab.

We walk a long way with the water at our shins. Suddenly Arnold grabs and juggles and has an octopus in his hand, pinched hard at the base of its body, fore and middle fingers held like a vise so it can't bite him. It's one of the daytime octopi, which is odd since it's nearly midnight, so perhaps they're more flexible in their habits than the textbooks say.

Its body is black and is the size of a large plum. Arnold has it held but has no control over the thing, any more than you would have over a splay of animated taffy. It is all over his hands; tentacles fine as spaghetti wriggling into everything, twining in his fingers like runnels of brown sauce. I put one finger into this melee and an inch and a half of tentacle tip runs over it and I don't feel a thing, I wouldn't know it was there if I didn't see, but when I try to pull away it has a hold on me like flypaper.

"Two of his arms are short, see? Just starting to regrow," Arnold says.

I don't see. Then I do.

"Probably a moray eel," Arnold says. "Morays like octopus. This guy was lucky."

We let him go, or he lets us go, and he blows off straight through the water with his arms behind him in a hank. Once he touches rock, he curls his tentacles under himself so that he looks like a toad and goes mottled and warty so that he looks even more like a toad. His big eyes are slit, like a goat's, and are set high and goatlike in knobs on the top of his head; he looks from one to the other of us and he seems about to speak to us, to tell us something, or lunge. He knuckle-walks down the rock in flowy bursts, changing his mottlings constantly so that now he is whitely warted and then browny smooth again.

Even after I have seen a half dozen octopi underwater at night and at close range I cannot get over how strangely made they are. They react to me with alarm, thinking the same thing, I suppose. They stare; they almost convince me that I am peculiar, that eight arms is the proper number, that jet propulsion is the only way. They always look me in the eye; they look me up and down, curled defensively on the surface of a rock. They don't run away.

I watch that pulsing jet of water-breath, which looks like a raw aorta, right where a carotid should be. Those two eyes, staring; I stare back. That big-nosed face looks like a trunkless elephant head but is really a belly. There's a beak between their legs and they can bite; I don't forget this.

Night is when their prey is out and about, or as out as it gets. They wriggle tentacles into reef holes and yank out shrimp and crabs, biting them between the eyes to stop them snapping. They'll envelop cowries or mussels, rasping holes in the shell with their radulae and injecting poisons and digestive juices, much as a spider does to a fly. When a moray attacks, they'll loose a cloud of ink that blinds the eel and overwhelms his sense of smell, like a skunk's spray would a dog's.

Their color changes are their language. After a while I know that furled arms and flashings of color signal alarm, but arms and flashings can mean other things.

The skins of squid and octopus are packed with pigment-filled cells called chromatophores. These are under nervous control, flexing directly from the pulses of the brain. The chromatophores are shaped like a tangly spider or a blot of ink splashed on a page; at the center of each one is a core. When the pigment spreads from there and fills the branching whole, the color shows. When the branches contract and the pigment retreats to the core, the color is hidden.

Hawaiian Night Octopus
Octopus ornatus, he'e puloa

We have nothing remotely like it unless you count blushing, which I do and always have done with very little provocation, so I think about it like this. A complex blushing in shapes and colors. This is what an octopus does.

Different species of octopus have different colors and patterns, which stands to reason. They flush bars and blotches and spots; they can hide or exhibit a palette of browns, grays, reds, yellows, oranges, even whites. With the contraction of skin muscles they can make themselves warty or smooth, in a kind of magnified goose-bumping. With this vocabulary they signal identity, alarm, attack, and sexual arousal, even in the dark — and they see well in the dark — and they can camouflage themselves almost anywhere on anything.

Their language for each other is a play of color and motion, something we could learn to read ourselves. And the odd thing is this: those chromatophores have quite a life of their own. Somehow this still shocks me, deeply, even after I have heard it said and have read it several times: chromatophores have a life of their own. This is what I saw in that Italian night so long ago: even after an octopus is dead its colors will change, the chromatophores will pulse, the colors will shift like shadows on a windy day, bringing up patterns, talking it over, for as long as an hour.

Songs of the Heart

The jeep rattles more than you would ever believe and the back is filled with gear and me. It is late in the afternoon. We stop in a small town for batteries, cans of drink, chips, figs, and go on through rolling country that is gray and prickly with pineapple fields, neat rounded rectangles with dark earth between.

We're going to Mount Kaala in the Waianae range. John is driving. This is his expedition. Kaala is the highest point on Oahu and is home to one of the few — if stunted and boggy — patches of native Hawaiian forest left anywhere on the island.

We're going to look for cone-heads, bush crickets, Tettigoniids; whatever you choose to call them they are nocturnal absolutely and shy in the extreme. They are the subjects of John's research. We're out to get specimens.

Their generic name is *Banza* and that's the name we use, though katydid would be close enough if I weren't among scientists. *Banza* it is. They are native Hawaiian insects, survivors in an environment that has welcomed wave after wave of successful insect immigrants. For the most part, these immigrants have eaten or have shouldered aside the natives. The *Banza* are still here. Odd.

Odder still, every island in the Hawaiian chain has its own distinct species of *Banza*. (The smallest island, Nihoa, has the biggest *Banza* of all; a twist in the evolutionary plot that we turn and bang on over beer after beer in the Japanese bar we go to, hoping that after enough riddling and drinking the skein will unravel and light will shine. It does not shine, but it twinkles; perhaps this is a joke of some sort and will resist, forever, the sober probings of our small and only rational minds.) And — as if this business of size weren't riddle enough — every mountain range on every island has its own species of *Banza*. Oahu, here, has two mountain ranges and two *Banzas*. Up the ante.

Banzas are flightless, which is why this is so. They get from island to island (or mountain range to mountain range, same thing) by tempestuous accident. This happens only once in a great while and in between great whiles our accidental *Banza*, dropped in new montane treetops by a fickle wind, sets up house and evolves into variety, race, then species, by degrees.

Their worlds are small: islands of montane forest in islands of sea. If one wanted to unravel the workings of evolution, who could think of anything more elegant than this?

First, however, you have to find the *Banzas*, and John is nervous. The pineapple fields go on and on. We come to Wahaiwa village; wooden storefronts, splashes of groceries and gas stations and bebop record joints, then we're out into fields again and somewhere ahead of us is the sea, you can tell; the ground begins to shoulder down as if to dump us off. The island is small, I feel this. To the left are canefields and the dark rugged rise of the Waianae. The slopes go abruptly up into steep scarps clothed in dark green like a cheap nubbled cloth. At the top of the mountains, clouds boil around.

The canefields are as pale as new grass. We turn up into them on a narrow road of much-patched tarmac that makes the rattles (if possible — it is possible) more deafening than ever. We roar over a cattle guard and past a herd of steers on the rough rising slope and past three horses behind a wooden rail, then we begin the climb in earnest in curves laid on curves like the coils of a rope. Suddenly we're up in strange thin woods through which there are glimpses of green twilight air; pure space plunging away.

It is too noisy to ask questions and my questions are piling up: what is that, what is this?

What do I know? I think. *What do I already know?*

Insects are old, old. They were already winged and plentiful when the first four-footed half-amphibian thing lumbered up out of the sea and breathed air.

Cockroaches carry on with one of the most ancient and pure of insect designs, and crickets and katydids and grasshoppers are closely related to them; they are all Orthopteroids, part of the same insect order.

Wings; that's what they had. Insects were safe in the air until birds developed from a line of agile little insectivorous dinosaurs, clambering leapy little dinosaurs lightening and feathering as they went up and up and finally out into flight itself. Superlatively eyed and fast and, now, winged, birds won that battle, if not the war.

Fossil evidence says this: that during the Mesozoic when birds evolved (and they evolved fast) there were massive extinctions of insects. Whole insect orders seem to have disappeared almost from one year to the next.

There is no good explanation for this except for the arrival of birds. Birds ate insects into oblivion whenever they could. Any insects that survived the predatory onslaught had to find evasive strategies and find them fast.

The strategy that worked best was to go where birds could not follow: into swift flight, tiny sizes, cryptic colors, crevice and earth, and — more than anything else — into the night. The insects that survived did this. Not that it was easy. Broad strategies are one thing but the details take some working out. Look at what's here: crickets and katydids are either superbly camouflaged or nocturnal, or both. Female crickets and katydids have a long curved ovipositor for drilling into plant stems. They lay their eggs in the dark hours, deep and out of sight. They have acute taste/smell sensitivities so that they can find the right kinds of stem to lay their eggs in. Each species feeds on different plants and it's important that the eggs are laid just in those. The Banza feed on grass. Their eggs have never been found, at least by people, and this is a tribute to their ingenuity as much as anything else; but it's likely that their eggs are laid in the grass stems. The hatchlings pass through half a dozen nymphal stages before they reach adulthood, shedding their skins between stages; these Banza nymphs are green. Grass green. At dawn they stretch themselves along the grasses that are their food, flattening themselves belly to stem and laying their long antennae — their antennae are longer than they are themselves—along the stem's axis. They stretch one antenna forward and one back; all the better to blend seamlessly in and keep in touch with their world. In daylight they are invisible. Only after dark do they move, or eat.

Once they're grown to adulthood and are ready to mate, there is another problem: how do they find each other in the dark?

The answer is that they have instruments with which they make songs. Adult male crickets and katydids and grasshoppers all fiddle, or stridulate, scraping the serrated file on one wing over a sharp edge on the other; this is like rubbing a knife edge hard along the teeth of a comb. The wings resonate and amplify the music like the skins of drums.

Males and females both have ears on their first pair of walking legs, near their knees, and what they respond to is the song of their own species. The song of a male Banza from the Waianae is distinct from all other Banza songs, in syncopation and in pitch. Hearing is vital: a deaf male will lose touch with neighboring suitors, will scramble that alternation of chirps which passes among them for the language of aggression and territoriality; he'll mess up the niceties of competition. A deaf female will not mate at all. The hearing of her male's song is her ultimate stroking of instinct, loosening hormones and keying her eager and patient advance.

Tonight we will find them as they find each other; by listening for rhythm in the dark.

We stop just at sunset on the final narrow pass to Kaala. Peaks as sharp and dark as blunted spires are around and below us. Far away and down is a swatch of pale sea and a level spot of green the size of my hand, with toy houses and bumps of palm trees. We are perched up on ground that seems to be too thin and sudden; the pass has the silhouette of a piece of cardboard held on edge under a blanket. Cloud blows around us like a thin shaking curtain. In the eddies we can see the beads of white that make the cloud. We can feel them on our skin. We are cold and wet; we put on sweaters and get in the jeep and go on.

Night comes quickly and cloud closes in, so that when we stop on a piece of road it is like stopping nowhere.

We put on our rainsuits and boots. I put four vials in one pocket and a notebook and pencil in the other and load a flashlight with fresh batteries. John nods; I follow him into the mist. He is carrying a white bucket of gear and I follow its pale bobbing shape.

This is like wading in cold and sodden wool. This is like no time at all, like no place.

While we are walking, our rainsuits make too much noise to hear much else, so we stop now and then to listen for Banza. We walk along a narrow little path through the bush where no one has been for a long time. The ground is boggy and soft and our feet leave deep prints that fill with water and black mud.

Once we stop by a draggled little bush that looks as if it were dying, scrabbles of branches with tips of leaves still thick, firm, glossy green, white furred beneath. It has three flowers; shaving brushes of bright red. This is ohia; the famous native Hawaiian ohia of the cloud forests, a plant that blooms all year long, that can take the form of tree or shrub, conforming to biological parameters no one knows or understands. Like all the other trees and shrubs here it is clothed in moss as if it were wearing socks; moss with ferns growing in it, with carabid beetles hunting through, and festoons of spider draglines.

Sometimes when we stop we turn our lights off as if we could listen better without having to see. This works well; tuning out one sense the better to tune another. With lights off we are truly afloat; there is no sense of sky or edge, or even of height; my feet have no solid purchase in the moss and jellylike ground. It is a dreamy blowy place with bumps of shrub, with the bones of dead trees rising up, and *olapalapa* leaves blowing and tossing. The *olapalapas* look like cottonwoods, they have the same spade-shaped leaves with flexible petioles so that the leaves flip and seethe, letting the wind run through them, but they are not related to cottonwoods or to anything else I know. The Hawaiian name sounds the way they look; none of the trees up here is remotely related to anything familiar, so that even in the dreamlike dark, with the mist sluicing through, I know that I am far away; for the first time I know in my bones that I am far away, with nothing but a few flecks of land between me up here and Japan, or Mexico.

For the first mile or so no Banza are singing. Nothing else is singing either. I remember then something I read about a Panamanian katydid that doesn't sing at all because its songs attract the predatory attention of bats. The male Panamanian katydid dances instead, the tap-dance vibrations carrying his seismic tune and intention through branch and leaf. The female gets the message through the soles of her feet. How much goes on here that we would need hair-fine seismometer feet to understand?

John is tall, lanky, fine-boned, with a taut nervousness in him so that I never quite know how he is going to move or what he will say. He thinks it's odd that there are no Banza calling. Last time he was here there was one buzzing every thirty feet along the trail.

A few nights ago I went with him to a hill above Honolulu to hunt for Banza. The hill was clothed in a riotous tangle of escaped fruit trees and vine tangles, and a narrow black path led up and through, slippery with wet. Here the Banza had colonized alien forest and John was excited.

"You hear that? Hear it . . . there!" he would say, stopping on the path, but I heard nothing that I could separate from the lacing of branches, the wind in the dark.

He explained to me then that the songs of Banza and other cone-heads are so high that most of their notes are up beyond the range of the human ear. At best we can hear only the lowest end of their music, a piece of the tune, static and faint. As we get older, he said, the high end of our hearing naturally diminishes. Our sensitivities close down gradually and without our really noticing, so that some spring, some summer, we go deaf to the piping in the trees. Cone-head research has to be grabbed at while you're young, he said. When you lose it, he said, you have to turn to crickets — crickets sing at a lower pitch — or cockroaches or walking sticks; or some other Orthopteran more amenable to our loss.

I wondered then if I hadn't already lost what I never knew I'd had. Later I thought I did hear one, the faintest of scrapings, more vibration than sound, so tickly and soft that I couldn't define its direction, though I could say:

"Yes yes! Now . . . now!"

"Yup!" John whispered. "Right over there . . . here goes nothing . . ."

He clambered into a tree. The tree grew over a ravine and

John crept carefully out and up in the branches, all long legs and long arms, the moonlight reflecting from the round panes of his glasses; I thought then that he was as much like his quarry as it was possible for a man to be.

At last he came back in and down to the path with the insect in his hand. This male Banza was longer and finer than any cricket, with a small pointed head, little beads of eyes, thin angled legs, immensely long antennae curving and tapping delicately, curiously, like silken wires. Its body was soft cocoa-brown marked with red along the abdomen. From their last molt the green juveniles emerge in this milk-chocolate vestment, the better perhaps to hide themselves in daylight against the tree branches where they sing. He was a handsome, calm little beast. We put him in a vial stoppered with cloth. John keeps dozens of Banza

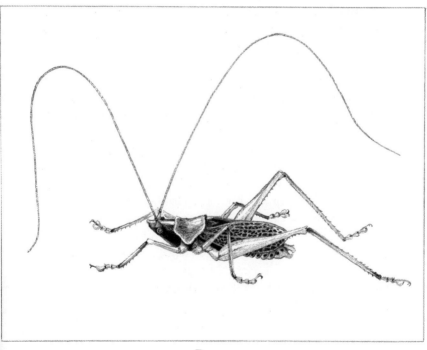

Banza

back at the lab, in glass boxes, and they are easy to feed and keep
while he observes them and does his research.

Now we are after more. We wade through the foggy forest full
of stunted tree ferns and shrubby ohias and other plants with the
sad leaves and poky form of potted avocados. A half mile in, we
stop and take the trapping gear out of the bucket and hang a
white sheet in the shrubbery and arrange a black light against the
sheet, fooling with the wires and batteries until the light is on
and the sheet is a pale purply-lit square. Insects are attracted to
these borderline wavelengths and they will come and cling to the
sheet. This is a good way of sampling what is up here and silent,
but this is not a good night for collecting. There is too much
wind. Insects don't like to fly in the wind. They lie low. But
something may come.

A little farther on we both hear Banza at last. We stop as if
we'd been hit; the soft creaking . . . there it is! And there! Two
Banza are calling not far from each other. Just beyond there are
four or five more. They call in turn, each challenging the other
and laying claim to his own place. They are deep in the scrub.
When we try to sneak toward them we make noise and they go
quiet. We stand in the wet bushes waiting for them to sing
again.

This is what we spend our next hour doing: stalking Banza.

I can hear them just fine. I'm pleased with this. I sneak up on one
while John sneaks up on another. I turn my light off and move
into a patch of head-high brush by the trail just where I hear the
singing, then stop and wait for it to begin again before I move
closer. Every time I move it stops, when I stop it waits; sooner
or later it comes again, closer, just there . . . no. There, ahead,
the height of my chest, no, lower, no . . . to the left a little. So I
go to the left a little, scraping against branches and getting a
spout of water in my face from stirred leaves, and the rhythmic
creaking stops as if it had been cut off.

Finally I know I have come within reaching distance. He's at shoulder height and now his ratcheting fills my braincase as if I were plugged in to some ethereal current, though only part of this comes in as actual sound. It gives me a clue to the real volume of his song: no whispers in the dark, this, but a body blow of noise. Who could resist? I stand and let it in. I feel unbodied, flung open by sound. This speaks wordless things, the way music does, or the repetitive pounding of a chant that joins the rhythm of the blood and sweeps the senses off. What is this? For minutes I stand in a bush on a mountain in the wind and dark, stunned.

Then I turn on my light and aim and there he is, on a branch, the color of milky coffee; his antennae dance in my light and I've got him.

Afterward we go back along the trail to check the black-light trap. John says that there have been nights up here when the whole sheet was paved with the shiver of wings. This is not one of those nights, we don't expect it to be, but in spite of the wind we have pulled some things out of the dark. We crouch there looking as close as we please, forgetting the wind and wet altogether, forgetting the Banzas in our pockets. But something stays. The Banza song seems to have sizzled the joints of my bones; I feel loose, afloat, half in a dream. The purpled surface of the sheet is studded with color; there are two kinds of sphinx moth, one in furry grays with white and hot pink along his abdomen and underwings; the other all airbrushed browns and yellows, elegant. There is one huge Noctuid, the smoothed shape of a Stealth bomber, with a furry slim-mouse body and a wingspan five inches wide. His wings are patterned in spots and waves of soft yellow browns, creams, coal blacks, scatterings of baby blue and orchid. . . .

All the moths shudder as if they were drugged, hanging to the sheet with their feet, which hold invisible rows of claws. The moths are not all; the mountain night has been sifted as if with a net. Here is a slender black wasp moving his abdomen like a warning obsidian finger, his wings held up like crumpled cellophane. Here are quantities of tiny flies, the flies that will be caught by the tiny spiders of the moss forests, the spiders whose mist-beaded threads make every branch glitter as if they were laced with silver wires. Here are little moths, a half an inch to an inch wide, some with wingspans like butterflies, others humped and woolly, others almost linear, like arcs of fluff.

These moths have to find each other in the dark, too. Moths have ears of a sort, on their abdomens, and some make courtship clicks and clucks, but most of their night songs are not only inaudible, they are not even sound. They are songs of scent. They are songs of time. They don't belong in this chapter, just as the sphinxes and the Noctuid — so furry and colorful, shuddering on the violet-lit sheet on Kaala — do not belong. None of these are Hawaiian moths. They are aliens, brought in as larvae in the wood of a shipping crate, in the rind of an imported melon. They have stolen this ground. I know this, but it hardly matters, I think, because they are beautiful. I'm not enough of a purist in these things, not yet. When we turn off the light and shake the sheet in the wind, they are gone.

It is early in the morning now, more than halfway to dawn. Back at the jeep we strip off our rain gear and pack the Banzas' vials in cotton wool for the trip down. We set off; I am curled in the back among piles of wet slickers, muddy boots, boxes and tangles of gear, buckets; there in the infernal racketing of the jeep and the grinding of its unhappy gears I fall deliciously asleep. I have wonderful dreams full of softness and color and exquisite melodies, and I am rude to John when he wakes me up at my house because I don't want to leave the dreams. He has to shake me and shout at me to make me move.

The next day I go to his lab to thank him and to apologize. He is sitting at his desk and his glasses catch the window light just as they did on the hill in the moon, while I try to explain my ridiculous dreams.

"You don't need to tell *me*," he says. "I know all about that. It'll do it to you, it'll make you crazy every time."

"What will?" I ask.

"Love," he says.

"Whose? The Banzas'?" I ask.

"Anybody's for anyone. Sure, Banzas' too, why not?" he says, and laughs. "In the end, you know, it's all the same."

Dances

I've wondered about the evening swarms of little flies, but only in bits, in one of those out-flashes within a conversation when words drag and my mind goes off not even tangentially, but wholly off so that the sentence I'm in the middle of trails away on the cusp of meaning, meaning nothing at all:

"But maybe . . ."

"It's as if . . ."

Incomplete. I'm staring at the flies instead of through them; seeing the fabric of a veil instead of the face beneath, the darkening trees or water. When someone else finishes speaking, I say, ". . . What?" a little too late to pretend I've heard.

This is about dusk, not night. This is about entrance, not inside. This is the veil one sweeps away or through. Twilight: the plummeting of candlepower, the key in the lock.

I've seen fly swarms here under the banyan trees, at home, too, and on a path through the dunes on an island where I used to visit a friend in summer; I realize now that I've seen them almost everywhere. They form up just at dusk in attenuated ovals or ragged spheres inhabited by rhythm so intense that one stares awhile without focusing on anything much, bemused by sheer vibration. One can walk right through it and feel no more than tiny taps as if from a light rain, or nothing at all. The thing sways, resolves, and goes on dancing.

As the darkness deepens they disappear. Or do they? I've never thought about it much. It's one of those things.

Once you've begun to unravel the illusions of the dark, you have to go on. Every mystery has meaning and it may be that the meaning is wonderful; most of these dances, after all, are dances of courtship. I've read that some flies dance over known grounds — a grassy hummock, a sheen of water — now, taking nothing for granted anymore, I wonder: in what language is the whir? Is it sound they're giving off, chorusing, or a scent that wafts, or is it just that jazzing vision that attracts, that drives the ladies wild?

It's the dancing of flies I'm after now, and I've come to the right place. It isn't that flies don't dance everywhere; but here some of these dances have been watched, deciphered, and understood.

And not just the dancing. That's only half. If a male fly's dance is a formal wooing then there's another half to the business, too: the dancer wants to be accepted by a mate. This is the purpose of the exercise, after all. This is the crux: the to be or not of yielding, of acceptance. The problem of whether or no it will be *yes*.

I always thought there was something, you know, that was distracting. I was distracted.

Distraction is recognition: the prelude to hunch.

Much of science consists of codifying hunch, of passing a nebulous wondering through precise prisms; of pouring the white light of living through and through until an array of data emerges in clear color; weighing the clarity in the hands, in a certain mood, saying: can I be sure?

The questions here seem simple enough: how are mates selected, who selects them and why, what does this do to the offspring in general and the species in particular?

The questions may be simple but the answers, as usual, can astound. For starters, it has turned out (not to everyone's surprise) that the selection of a mate is almost always and everywhere a female prerogative; among whales, panthers, birds, snakes, mice, even — perhaps most magnificently — among flies.

The swarms gather and whir under the banyan trees, beside the waterfall, over a stone, and in other places less visible to us, and the night comes; in the light of morning the world has changed and life has changed in it. Mating precedes the interscramblings of genetic material, which affect the mating dances of the next generation; so it goes, generation after generation, nuance deriving from nuance, until one dance has splintered into two or four or sixteen: sixteen separate dances and sixteen species of flies.

The conclusions were tentative once but are gaining momentum; Darwin didn't have all the answers. Sexual selection has a power he never knew. Mating may be the greatest and riskiest of knots tied in the world, but it now appears that it is also one of the Great Knots on which evolution swings — to the everlasting glory and diversity of life.

Perhaps it's a sign of the times. The dance and the yes. What are they and why?

The sun reddens and the mountains darken; the slopes change shape with the subtraction of their light, gathering dusk into their folds. I pack my gear, my black sweatshirt and tape recorder and notebook; it's almost time.

Tonight I'm going to see the dance of the melon fly, *Dacus cucurbitae*, a honey-colored housefly-sized creature whose larvae feed on and happily ruin all sorts of melons and squash. The melon fly is an unlucky import to Hawaii, and here as elsewhere it's a pest and a cousin of pests: near brother to the oriental fruit fly, relative of the medfly, part of the dreaded brotherhood of agricultural incubi. Whatever we think of it, the melon fly is, in and of itself (and not without help from us), an enormous success.

In the case of the melon fly, success translates to the helpless pulping of watermelons, bitter melons, cucumbers, canteloupes, gourds, butternuts, pumpkins; the manna in which melon fly larvae hatch and feed.

Watch the flies. You would never guess that deep in the arcana of the mating dance of the melon fly lies the clue for controlling its pestiferous nature.

*   *   *

Someone has guessed just this. It's more than a guess; it's been through the prism and come out clear. He likes to be called Ken; to his colleagues and students and associates he is awe-inspiring, a man worthy of any number of his hard-earned honorifics. Ken Kaneshiro is a compact handsome man with thick salt-and-pepper hair and eyes that miss nothing. He is a world class entomologist. Nowadays he is the director of the large and prestigious Hawaiian Evolutionary Biology Program at the University of Hawaii. He has spent much of his time there studying the intricate mating behaviors of native Hawaiian Drosophilids, a wildly diverse group of native flies. Now he's studying the melon fly, too. It happens that they have a lot in common.

Just at sunset Ken comes to fetch me, and we drive into the country. The air congeals over the pineapple fields in a lavender mist, the hilltops are gilded, the borders of the sky are the clear translucent yellow of honey. We pull off and stop where there are crude fences around garden plots and ragged trees, a shabby place in a hollow of the landscape. No houses are in sight and just the one road. It has a rural bleakness graced, now, with the richness of the light and the excitement of the chase.

I follow Ken into the trees on paths that go between the gardens. I have to run to keep up with him; he's lithe, fast, with the attentive grace of a hunter. One gardener is still washing off his tools by his shed, getting ready to go home. He waves at us and salutes Ken: he recognizes the distinguished scientist from Honolulu who comes to the gardens at dusk in his black boots. Soon we will be the only people here.

The color is draining from the trees. We hurry along the narrow paths between rectangular plots surrounded by bamboo and wire. These are public gardens, each one taken by an individual or a family for their own use; there are banana trees, papaya and guava trees, patches of taro, lettuce, onions, beans, melons, radishes. Vines of winged beans and green squash clamber up the fences.

We come to an arbor over the path. The vines on it are dead and the leaves are shriveled.

"This was one of the best places," Ken sighs, "but I can't stop them from cutting the vines. Come on."

We're off again.

When Ken began to study melon flies, he wanted, first of all, to find out all he could about their mating behavior. It turned out that no one knew anything about the mating behavior of melon flies in the wild, and this is odd when you think of the hundreds of thousands of dollars and gallons of pesticide loosed on them over the years; but no one had. No one thought, I suppose, that the mating behavior of a fly could be of any importance. If people had seen it they did not realize the crucial nature of what they had seen, but this isn't so surprising when you know, as Ken does now, that the dance takes place on the undersides of leaves, in the dark.

We go on, looking at leaves as we go. I find one female sitting on a bean leaf, the pointed honey-gold abdomen (all the better to drill through squash rinds) and the gleaming wings spread serenely out. She doesn't move when I poke the leaf.

We look on and under other leaves nearby but find no others. Ken shakes his head and we're off again.

Ken has discovered that melon flies gather to dance and mate in certain places every night. A gathering place — like the arbor of vines — may be used every evening for months on end. This was not unexpected; his Drosophilids had been doing the same thing.

In sexual terminology a communal mating place is known as a *lek*. Some birds use leks — prairie chickens, some species of grouse — and with birds or flies the pattern at the lek is roughly the same: the males come to do the displaying and the females come to do the selecting. The competition, naturally, is intense. This is the point. On any one night here on one vine arbor there may be five interested melon fly females and more than two hundred willing males. The males' performances are rigorously judged: forty to fifty percent of melon fly males never mate at all. Thirty percent of them get more than sixty percent of the action.

I think of the nickname and pro-forma work name of Ken's line of research at the university: "Studs 'n' Duds."

That's what's here.

There is a smell of evening earth, and a vegetable smell, part rot and part bloom. Neat rows of onions spike from the ground. The leaves of banana trees hang in the windless air, corrugated planes that reflect the fading pallor of the sky.

"Ken —" I say, hurrying after him, hurrying to understand, continuing the conversation that has been going on now for an hour in fits and starts; "Ken, you mean stud-ishness is a heritable trait?"

He nods without stopping.

"And Ken . . . and the choosiness, the females' choosiness, is heritable too?"

"Yes," he says, over his shoulder. He looks under leaves, ducking, peering upward; squash leaves and bean leaves are the ones to watch.

"And if you breed the choosiest females with the stud-iest studs," I continue, "I know I'm not saying this right, but if you do — then you'll get more studs. And if you sterilize those and let them loose —"

"Yes," he says, stopping and turning.

"— you can control the pest population?"

"Yes. That's what we think, yes."

"Why hasn't anyone done this before?" I ask.

He shrugs. "The USDA has had a biological control program for more than thirty years," he says softly. "The problem is, the male flies that they're sterilizing and releasing are strains that have been bred in laboratories for many, many generations." He pauses, letting that sink in. "The strains have been selected, mainly, for the ease of breeding vast quantities of themselves in laboratories." He lets that sink in, too. "When those males are released into the wild, how do you think they're going to find the leks? How do you think they will compete with the local males? Well?" He laughs.

"You mean, they're releasing a lot of *duds?*"

He nods, smiling. He represses his smile. It doesn't do for a scientist to gloat, ever.

"The really exciting thing," he says now, in the near-whisper that dusk inspires in anyone, "is the opportunity to apply twenty years' worth of basic research to an agricultural problem."

Which implies that excitement is where it's at, for him. I don't doubt this now.

We're off again. A single bird hurries overhead, a black shape going into a tree. Birds are settling for the night. Now it begins: insect time. The thickening honey-and-rot smell of earth and ripe melons and overripe melons. This is a garden anywhere, a garden everywhere going into the darkness.

We stop under a guava tree covered with vines of a wild bean.

"This is another place," he whispers.

"Another lek?"

He nods.

He doesn't expect much; it's winter now and the fly populations are low, but here they are. There are three . . . no, five, eight . . . ! It is easy enough to see their silhouettes against the leaves. The sky is the neutral color of calm water and the leaves have the translucence of jade. There are more flies now, more. Dark crosses on the jade.

Here is a female on the underside of one leaf. Her wings are spread and her abdomen is pear-shaped with a stiletto end. She is absolutely still.

Here are two males, they are on other leaves. Once in a while, they move. Then they move again.

Now and then the males fan their wings; the black markings in the wings' transparency appear and disappear as they fan. Ken says:

"They are fanning pheromone. Aphrodisiac scent."

I nod.

"The wing vibrates against a row of bristles on the abdomen," he continues. "The vibration atomizes droplets of pheromone that are excreted by the tip of the abdomen." He pauses; we watch. "The pheromone is atomized into a cloud that surrounds the wing-vibrating male," he whispers. "There . . . see?"

I nod.

We watch.

It's like a ballet; the performers coming in from the wings, all the evocative and gentle beginnings, the orchestra tentative — all strings and one oboe wistfully tooting low notes — the prima donna poised, central, waiting to be drawn from her virginal trance . . . already you know how it's supposed to end, in ecstatic pas-de-deux, but you sit there glued to your seat.

One male lands on a leaf with another male. He fans and fans, then leaps to mate; the other male flies off.

"Toads do that, too," I say, under my breath.

"What?" Ken says without moving his eyes. The thing is to keep one female always fixed in your sights. I'm doing this, too. My neck hurts.

"Male toads, in mating time, will grab anything," I whisper, "Mud lumps, other males, anything that stands still." I bite my lip to keep from laughing: I haven't seen anything funny in this before, but there it is.

Ken smiles. I see his teeth gleaming in the dark.

There, again; a male lands on a leaf with another male. My fists are bunched. They fan their wings at each other and they both fly away.

The female waits on her leaf.

I think of what Shika said, last night. Shika is a research associate of Ken's. She is the curator of the insect collection at the Museu de Zoologia in São Paulo, Brazil, and is known in official circles as Dr. Francisca C. do Val. She did not — at two in the morning with her bicycle shackled to the lamppost outside and barrettes in her hair — seem professorial. John and Aubrey were with us and we were drinking beer.

"When I begin with my students," Shika said, leaning forward, "to discuss with them the choosiness of the female flies, the differential in female discrimination that corresponds genetically to Ken's studs 'n' duds, the women in the class always begin to write on a little bit of paper." She grinned. "They try to find out, you see. How choosy or not choosy *they* are." She laughed; John and Aubrey looked nonplussed. "On what basis do we choose?" she said, "That's what the women are thinking. You see! I know . . . you are thinking of it right this minute! How do I choose! Right?"

"I never thought about it before," I said.

"I think about it all the time," Aubrey said.

"Women —" John began.

"You see?" Shika laughed.

Now maybe I am going to see.

There! A male lands on the leaf with one of the females. There he is.

This is it. He comes toward her at an angle, his nose to her flank. He fans his wings, in rhythm, first in short bursts, then in longer bursts. Ken whispers:

"We have held a microphone to that, and we have found that the males are making noise with their wings." He pauses. "And they sound like moans: 'uuungggh . . . uunggh . . . unggh.' "

I watch. He watches.

Then he whispers: "And you can see, they go faster and faster: 'bzz . . . bzz . . . bzzz . . . bzzzzz . . . bzzzzzzz . . .' until you hear a continuous 'bzzzzzzzzzzz.' They build to a crescendo." He pauses. "Only it isn't 'bzz.' "

The male closes in on the female in tiny steps, tiny steps, fanning and fanning all the time and then he leaps but he leaps onto air: the female has flown away.

Ken laughs.

"Most of the time it ends like that. She flies away."

We watch others.

Ken tells me that he has watched melon flies courting and mating (and not mating) in the laboratory, too. He noticed there that once melon flies did mate they stayed together all night, attached, in coitus, until they were touched by the first light of the sun. Then they separated and flew away.

He once kept a pair in a darkened room to see what would happen. They stayed linked in the dark for three days and nights until they died. Ken seems embarrassed by this, almost apologetic, though it told him something he needed to know and it was only two melon flies, after all.

While we watch I ask questions, and Ken tells me how it is that these competitive dances, these songs, scents, fannings, these choosy choices, can lead to the evolution of species. Sexual selection: *change* is what it means. I think I'm beginning to understand. Then he says:

"The basic concepts are the same for all organisms. The basic research findings can be applied to all organisms."

"John said that, too, in a way," I whisper at last. "And Shika. She said it, too, in so many words."

"What?"

"That love is the same."

"I'd have to say," Ken says at last, "that they've got a good point there. Of course . . ."

There's a pause. The flies are darker crosses on darkness, still dancing. There is a star out now above the mountains.

". . . there are differences," Ken says at last.

"Yes."

"But maybe . . . ," he says.

Another pause.

". . . the differences are not as important as the things that are the same."

Ephemeral Eternal

Creatures that spend their lives in caves are known as troglodytes. It's a word that has gotten under my skin. I chew it over

when I'm driving in the car or walking; repeating it like a charm
or a phrase of music, letting its message chunk home. Troglodytes: inscrutable, brittle, veiled in impermeable black.

Caves have no days or nights or cycles of seasons. Nothing
much changes there and nothing much comes in.

To be in a cave is to be in eternal and utter dark. If you sat out
in space as far as possible between galactic cores, you'd have
more light than this. If you happened to bring along a telescope
and a slide rule and a pencil and paper and Newton's *Principia,*
you could figure out that time passed, you could know where
you were; but caves have no points of reference. None. Nothing
is there except for gravity, the ticking of the heart, hunger, the
cyclings of urge. Troglodytes live night lives in the extreme.

I always intended to write about life in caves, as an exercise in
nocturnal ultimacy, and I thought of the limestone caverns of
the Southwest or the Mid-Atlantic; I thought I would do something there. I thought of stalagmites, the echoey drip of water,
blind translucent fish and white crayfish and eyeless salamanders
in crystal pools, an overhead paving of bats like knotted velvet. I
did not think of Hawaii.

Hawaii has caves called lava tubes. What little I read about
lava tubes did not prepare me for these; they were mentioned as
a geologic anomaly, a fluff, a peculiarity of lava flows; lifeless
punctures, of interest to no one except, possibly, to volcanolo-
gists. They're one of those things that just happen. If you picture
molten lava pouring downhill, flecked with crusty dark like a
dragon's tongue, you can see it happen: a riverlike flow hardens
on its surface and encloses molten stuff like the sausage in a skin.
The encased lava river delivers molten rock to the front, over-
flowings raise the walls and thicken the roof; when the rock
congeals what is there is hollow casing: a cave.

Lava tubes are chancy things. Most of them crumble or sag
before they cool, others burst from the pressures of gases
within. If they last through the toughening of their rock and
escape being crushed by subsequent flows, their ceilings collapse
or erode soon enough, crumbling from the wear and tear of

nothing more violent than simple years of rain. It would be hard to think of any places more inhospitable to life than these: utterly dark, sterile from their birth, isolated one from the other by layers of lava and, between islands, by stretches of air and sea.

Troglodytes anyway tend to be isolationists. The nature of their trade has made the outside world as deadly to them as deep space is to us. They inhabit an archipelago in a sea they cannot cross. Birds or winged insects or plants may take to the wind and colonize where they fall, but cave life developing within one island of stone could hardly be expected to spread to the next.

All these things are daunting enough, but the most life-defying element of lava tubes is their ephemeral nature. They are environments with expected life spans of perhaps ten thousand years. This seems hardly enough time for a cave species to evolve on its own from whatever surface life would care to penetrate the dark.

All things considered, common sense would have it that these caves are lifeless. Until one day in 1971, people thought that Hawaiian lava tubes were just that: barren fragile bubbles soon popped.

Frank Howarth likes caves and always has. Cave exploration has been his hobby for years. Penetrating the ultimate in darkness has its attractions; what is in a cave is immune to weather and tends to remain there undisturbed. Exploring any cave is a voyage to the geological heart of the matter, a mystery tour. You never do know what you'll find.

Caves hold treasures. Hawaii's lava tubes are known for catching and protecting the flotsam of history: bones of extinct native birds, blown seeds of extinct flora, remains left by ancient Hawaiian people. Caves hold the present as well as the past; some certain cave may be as sacred to modern Hawaiians as it was to their ancestors: a portal of the subconscious, the orifice of sacred dreams. Even to the secular entrant the lava tubes are evocative and beautiful, with ropy lava-flow floors, talus heaps and lava boulders, "lavacicles" like stalagmites and stalactites of molten stone, endless varieties of frozen-flow configurations in utter black.

Frank Howarth is an entomologist by profession, so his eyes were attuned to the minuscule even before he entered a Hawaiian cave for the first time. Caving was, as I've said, a hobby for him, in Hawaii and elsewhere; but Hawaii was where he happened to be working and where he happened to be caving; and one day in 1971 he went off to explore a lava tube on the Big Island. He wasn't looking for anything more in there than a nice day off.

Deep in and far beyond the reach of the faintest glimmer from the small opening, he shone his headlight on a curtain of ohia roots that had penetrated the lava ceiling. The roots hung there in a stringy tangle.

Attached to one of the roots was a planthopper, a sap-sucking insect. It was feeding on the ohia root. It was the color of the ohia root; pale, eyeless.

That was the beginning.

The planthopper took Frank by surprise. When he got over the surprise, he went back into cave after cave armed with a pair of forceps and a pocketful of vials. He came out every time with the vials full. The richness of Hawaiian cave life took everyone by surprise. If Hawaii is the "showcase of evolution" that it has proved to be (all its thousands of insects having evolved from three hundred to four hundred immigrant species that blew in by chance, in jet streams probably and from the east mostly), the caves are a kind of dollhouse showcase, environments in miniature, scant in size and brief in time.

These islands are ephemeral enough in the scheme of things. Hawaii, the Big Island, is just three-quarters of a million years old at its oldest end. Its newest end is growing as you sit, as lava creeps down the flanks of Mauna Loa and Kilauea and hisses into the sea. Kauai, the oldest and westernmost high island of the chain, is 5.1 million years old and eroding fast.

West of Kauai and little Nihau is a pearl-like chain of remnant rocks and flattened sandy islets: Laysan, Pearl and Hermes Reef, Midway. West and north of Midway a line of submerged seamounts stretches nearly to the Bering Sea.

Islands, reefs, and seamounts are phases in a process. They were all born as the flat plate of the Pacific basin drifted slowly from east to west over a "hot spot" in the earth's mantle underneath. The hot spot pumps magma pretty much all the time and once in a while the ocean floor ruptures, like a boil, letting the magma loose: a volcano. As the plate drifts westward the volcanic islands go off with it, cadging a ride. Almost before they finish growing they begin to erode back into the sea, their breast-like rounded cones first riven into peaks and gulches, then into choppy islets and atolls; finally they disappear under the waves.

This isn't to say that the islands are negligible things. Hawaiian volcanoes are the largest volcanic formations on earth. The base of the Big Island's Mauna Loa is 250 miles thick, the peak stands 33,000 feet above the seafloor on which this base rests. The topmost 14,000 feet are above the sea, but will not — in the geological scheme of things — be up there long.

Meanwhile Mauna Loa has its share of lava tubes. The caves are like froth in the volcanic bodies that made them.

I wanted to go into a lava tube myself. When I spoke to Dr. Howarth on the telephone and mentioned this, he replied with equivocations, silence, then with cautionings and polite dismay. It didn't take me long to figure out that this was like asking the curator of the Museum of Modern Art if I could take a hike across his Mirós and Picassos: caves may be fragile and unique by nature, but no caves are more fragile than these.

Lava-tube creatures may be adapted to darkness and dampness, to surviving the drift of sublethal volcanic gases, to finding widely scattered foodstuffs, to navigating on hard rock and slick surfaces and through mazes of crevices and channels; but there is one condition they cannot tolerate at all: disturbance. Disturbance is outside their experience altogether.

This is fragility taken to its highest degree. The web of lives that has woven itself in caves is like a network of crystal.

Of course, people do go in. According to Dr. Howarth, people go in too often. Hobby cavers are legion and lava tubes have been "discovered" by them and Dr. Howarth wishes heartily that they had not.

Caves, like alpine meadows and mountain bogs, can be loved to death, trampled to oblivion. There is a paradox here. If caves and their inhabitants are not made public, then they can be destroyed through ignorance; but increasing public awareness increases visitors; and visitors increase destruction. Knowledge about cave beings is knowledge at its most dark-angelic: a double-bladed sweep whose motion either way means damage.

Once I get the picture I decide that I will not go in. Knowing what I'd do there, I no longer want to go.

But I do want to meet Dr. Howarth. I want to see what he has found.

I am early for the appointment. While I am waiting at the museum where he works, I meet a scientist I already know. He's a colleague of Dr. Howarth's.

"You're here to see Frank?" he says.

"Yes," I say.

His voice drops to a whisper.

"Frank Howarth is an astonishing man, a great scientist!" He pauses then and looks me over, as if he is wondering if I can be trusted with so great a treasure. I wonder myself.

"You know, don't you," he continues, "that he's not only discovered dozens of new species, but half a dozen new *ecosystems*. There's not another person in the world who has done *that!*"

Suddenly Frank Howarth is here. I jump; he towers over me.

"Hello," he says.

Then he smiles, suddenly, as though he's just remembered to do it; but this is a real smile, not one put on for show.

He's a tall, attractive, graying man, and he moves quietly. When we go off toward his office, he seems to drift rather than walk, his voice is soft, and I notice that he has the hands of a watchmaker.

His office is too small. He assures me, with the kind of shy eye-twinkle usually erased in human beings by the time they reach the ripe age of six, that these compact quarters are temporary and that he doesn't really mind anyhow, that he spends as little time in here as he possibly can. The office has the feel of a closet: stacks of boxes, books, papers, bins of black-lidded vials, a desk snowed under by papers, two folding chairs tucked against a wall.

He takes a bin of vials and a pair of forceps from a shelf and we go to the library and the next two hours pass there in a kind of oblivion. We focus on the vial contents: preserved cave insects that are like tiny tangles of pale thread, knots of ivory. He handles them as if they were jewels, and jewels they become: brilliance plucked from a stygian mother ore.

There are thread-legged bugs, eyeless and pale as all true cave species soon become (evolution having a way of dispensing with anything, pigment or senses or nervous circuitry, that is unused or unuseful), and their legs are as long and fine as silk. They sit and wait for prey with their legs splayed like a web, their front legs mantis-like for the quick snatch. Their long antennae are sensitive to vibration, sound, and the warmth of infrared radiation; these search the "strike zone" constantly like a whiskery radar.

There are two species of cave crickets, both of them wingless and blind. One inhabits rubbly ground near the cave entrance, the other lives deeper in. They both have enormously long antennae that are sensitive to chemistry — to taste/smell, in other words — and these are enough to guide them to food. It doesn't surprise anyone that both species seem to have evolved from a lava-flow cricket; a cricket that lives on barren windswept mountaintops, feeding solely on detritus blown up from valleys below. The lava-flow animal is already deeply nocturnal, already wingless, already a scavenger in near-barren ground. One of the cave crickets still wears some darkish pigment, which shows how recently he's entered and taken up the life.

Another cave creature that seems to have come in from the lava flows is the hunting spider. The outside species is called the Big-Eyed Hunting Spider, which is an apt enough description — he looks like a woolly robot wearing racing goggles — and in spite of his enormous eyes he is made for hunting without much use of sight. He's equipped with bottlebrushy legs that are sensitive to taste, smell, warmth, and — especially this — the most delicate of vibrations.

The cave hunting spiders have no use for eyes at all. Like other cave beings they move slowly, ponderously. They lift each foreleg in turn, as if "seeing" ahead with the backs of their knees. They are as pale as chalk and their eyes are going or gone. Some caves on the Big Island contain the Little-Eyed-Big-Eyed Hunting Spider, three caves on Kauai contain the No-Eyed-Big-Eyed Hunting Spider; Frank found them himself and he named them himself. These hunting spiders are the caves' top predators, the peaks of the troglodytic food chain, the ends of the line.

The ohia roots are the base of the food chain. Pale penetrating roots are the energy supply for the whole cave system — these and the occasional creature that may blunder in and die there in the dark and the scant detritus washed in by rainwater. These are the raw materials on which the cave runs. Tree crickets and planthoppers feed on the roots; thread-legged bugs, a predatory earwig, the hunting spiders, and other smaller spiders prey on them and on the myriad scavengers; springtails, blind terrestrial water-treaders, centipedes. There are Noctuid moths, too, that spend their entire lives in caves. Like all troglodytes they have given up on wings and eyes and the outside world altogether.

The astonishing thing, the wonder, the head-shaking miracle of it all is that this rich assemblage lives together in caves that are very young. At least twelve entirely cave-living forms and up to twenty-five part-timers live in a single cave on the east slope of Kilauea, a cave that is, at the most, only five hundred years old.

Cave creatures may migrate underground from older to younger caves, so that some of these may have come in from neighboring lava tubes, but they cannot have come from anywhere else. Their ancestors may have been nocturnal, which gave them a head start on troglodyte technology, but they did not come from caves. They blew over here from another island — onto raw fresh lava, mind you — no more than three quarters of a million years ago. This is not long ago.

Blow on over. . . . It's like old-fashioned fire-fighting: pass the bucket. How long has all this been going on? Who can say? Evidence has collapsed, is collapsing, has washed away. For millions of years lava tubes have formed and eroded and cave creatures have, maybe, probably, who knows? — formed and eroded with them. But this much is sure and is wonderful: Hawaii, the Big Island, has at least twenty-three species of troglodytic insects. The island of Kauai has only two. What is newest is most diverse: evolutionary theory stands, here, neatly, on its ear.

An hour and a half has passed in the museum library and now I look around; people are coming and going, they're opening books and reading books and putting books back; I feel the pang of reentry that one feels coming out of a movie theater at dusk; from the regal to the plebeian, back to the daily dusty world of the merely known. Frank Howarth tweezes the glasslike specimens back into their bottles and twists on the tops. He looks at me hard, intently, as he has all along, as if to ask:

"Do you see?"

I see him going with infinite care along the narrow footpaths in the caves he has found, watching — for hours, maybe, because it takes hours — the precise and lugubrious mating ceremony of the Little-Eyed-Big-Eyed Hunting Spider. I see him plucking a single pale cricket to be looked at later in detail. He knows that spiders sensitized to the footsteps of millipedes cannot evade the bomb-tread of man steps, that tree roots torn mean worlds dead, that a planthopper nymph or a caterpillar once dislodged from the ceiling may never climb back.

He has been in lava tubes that are only a month old, the walls too hot to touch, radiating heat like an oven. Within his lifetime creatures that are the color of shreds of glass or chips of shell may come in here to live. New creatures, perhaps, new animals that no one has ever seen. If left undisturbed, they will live here for a few centuries or a millennium or two or ten and will move on to new and younger caves or will vanish and be gone. While they are here they will form an assemblage that will be unique in itself; each cave is entirely different from any other cave, host to a different community. They are islands within islands. They are unique within uniquities.

They are orifices of more than nightmares, the home of more than dreams. If you pass a cave mouth on these mountain slopes, half covered by vines and the scrambling green of ferns and the color chips of birds; if you glimpse under this beguiling veil the face of velvet dark, which is no night, because night passes in hours and this does not; if you remember that this is the world — cross your hands on your breast and bow.

Stringed Instruments

When I haven't slept enough, I think of strange things. Perhaps I am used to it now and I've learned to enjoy it; the skidding off the conscious road into a rural network of mental rabbit paths and old forest roads that are only just discernible in some luminous dark. The subconscious is analogous to night, in which nothing is precisely clear but everything exists. Neural threads arc out and splay into unimaginable distance as if the mind were infinite and curved: you can go anywhere, time travel not excepted; and suddenly when I'm packing my toothbrush I remember a game we used to play in the backseat of the car. We used to play it there all the time when we were kids and one of our mothers was driving us somewhere and we were bored. You think of a word and the other guy has to think of a word and then you do and then he does and so on as fast as you can go, a game of free association if there ever was one; a game of firing verbal neutrinos into one another's subconscious until critical mass is reached and the backseat of the old Ford wagon explodes with giggles.

I realize that I play this game all the time, without really knowing that I do; whatever I see or hear elicits a response: the backseat game of the world.

I have been worrying about perception. I am supposed to be packing my things now for a three-day trip to Maui but I'm gnawing over this problem all the time. It isn't just the upflux of memories that appear, they're an artifact of the central worry; it's that perception is never the world. Perception is what we make of the world.

For instance, sight is a wonderful sense. I can take this patchwork of color bits (wavelengths) and darker and lighter bits (relative intensities) and make a room and a suitcase on a bed and a window with Manoa Valley in it. No: I make a *model* of a room and Manoa Valley. It isn't the world. It's a model. The model is so good that I forget most of the time that that's all it is.

When you can't see much (and a lot of the time at night I can't see much), then you are forced to make a model in a different way. At night the world is a different place, and one is different in it. That living things make models in different ways is the central lesson of the dark.

I have a blind friend who carries a collapsible cane in her pocketbook. She can do remarkable things with that cane. She can get anywhere she wants in New York City just with that. Her city is built from tactile images; the clanging length of lampposts, indented rims of doors, ribbings of gratings, certain textured curves that may be curbs or the feet of friends. Touch: it can make a model, too.

*   *   *

Finally I'm packed and I've got my ticket and I'm on my way, and only when I am airborne and Maui is close do I remember that I am going all this way just to look at spiders. That's it; that's what jogged this loose; spiders know the world most of all by touch. Watch them: they tap tap forward and up and sideways continually as though they were playing on the world, as if the world were a piano and they were eliciting a symphonic response: a harmonic model. I shake my head to stop the spontaneous giggles as the plane sinks and the engine changes pitch — we seem to be descending into the sea — spiders are everywhere, after all.

But Rosie Gillespie is not everywhere, she is here waiting at the airport beside a van she has borrowed from the Nature Conservancy, and she is barefoot and wearing a cotton skirt, and she has blonde curls she never brushes. She greets me: "Diana!" in an English accent and with a smile. If she has come all the way from Britain to study spiders, then why should I think that I have come far at all?

She will show me the nocturnal spiders of the rainforest, and I've been told again and again that I could find no better guide. She looks like a beautiful girl and has a girl's softness, and only when she is tired can I see the faintest lines around her eyes that say that she may be past thirty, but not far past. I have heard her name uttered with awe in the corridors of the university at Honolulu, and by scientists twice her age, perhaps because she works three times as hard as anyone should and accomplishes ten times as much. Later, after I've worked with her for a day — a "day" that begins at three A.M. and ends at midnight — I know that she does not do this out of ambition, but because there is nothing else that interests her in the whole world.

She came here to study a tiny nocturnal creature known as the happy-face spider. The happy-face is a native Hawaiian spider named at two in the morning by a rudely awakened biologist, the rude awakener being wonderfully overexcited at his discovery (or rediscovery; no one had seen a member of this species for more than half a century and it bore only a Linnaean name:

Theridion grallator), and he was shouting "Hey hey! Look look look look!" so that his reluctantly roused colleague opened one eye, stared at a grinning clown's visage painted in nightmare colors on a tiny spider's abdomen, and announced: "Ah, yes. The happy-face spider . . ." and rolled over and slept, and remembered nothing in the morning at all. But the nocturnal christening stuck, and sticks still, and is apt.

The happy-face is in the same family as the black widow and the house spider: a successful and numerous family well spread about the earth. Like most spiders everywhere the happy-face is active at night, perhaps because birds prey on spiders and birds are sharp-eyed and can go almost anywhere that spiders live, at least in daylight, so the reason for the happy-face's habits may be simple.

Other things are not so simple. For instance, there is the problem of the patterns for which these spiders were named, but which they do not always have, and which (if they do have them) are rarely the same from one to the next. The happy-faces have globular abdomens smaller than a rice grain, and the whole spider is a basic translucent yellowy green like a seedless grape. Some happy-faces are just this: patternless their whole lives and none the worse for the experience. Others have patterns that look as if they were painted on by a pixie; a bored pixie, a pixie with nothing better to do. Some wind up with abdomens bearing two black spots, or constellations of spots, or red blobs, or black blobs, or red horseshoes, or yellow blobs with black spots, or red "smiles" with black "eyes," and so on. Including (which I wouldn't believe existed except that I saw it myself and have the photograph to prove it) a yellow blob-face inhabited by white eyes with pink pupils, arched white eyebrows, pink and white hair, and a red mouth curved in a perfect Cupid's bow. The spider that wore this abomination also (I see now, from the photo) had sixteen pink knees. An assiduous pixie.

This raises a number of questions, not the least of which are *why* and *how*. The answers to these questions could be nuggets of the purest genetic truth, if they are found, even though the finding may be akin to the process of passing a threaded needle through every blade of grass in a meadow. I look at Rosie as I have looked at other master scientists and see that this is life enough for anyone, more life than most of us ever attain.

Rosie is studying many things; not only the genetics of the variously patterned forms, but diet-induced color changes, the spiders' aggressiveness and how this changes with age, their movements, the abundance of their prey throughout the year. She is working on all of these things in the Nature Conservancy of Hawaii's Waikamoi Preserve, high on the slopes of Haleakala.

Rosie tells me about this forest as we drive. Waikamoi, she says, is more than eight square miles in size and is one of the largest patches of native Hawaiian rainforest left anywhere. It is unexploited and largely unexplored because it is difficult to get into.

One thinks of rainforest and one thinks: anaconda on the riverbank, tsetse flies, adders underfoot, panther overhead, leeches, mosquitoes, barbed vines . . . all these things. But here there are none of these things. Isolated as this is in the mid-Pacific, no reptiles or amphibians ever came here, and no mammals (except one bat) ever came here either. Only insects and birds. So: there are no snakes of any kind, no frogs, no panthers. No flies or mosquitoes or leeches suck blood: there has never been blood here worth the sucking. There are no grazers to crush the plants or eat their leaves; the native raspberry has lost its thorns, the native mints have dispensed with their repellent scents and flavors, the pea family has given up its toxic alkaloids. Even the birds are unafraid.

Most of the time it is raining in Waikamoi. The eroded volcanic face on which it lies is steep and riven. The forest is filled with bogs, precipitous slopes, deep gulches, surfs of neck-high ferns, impenetrable tangles. It lets one struggle through in silence, without a pang, without a thorn or a sting. This is where the happy-faces live.

Other spiders live there, too. During the many months' worth of nights that she has spent alone in the forest, Rosie has discovered at least seven new species of Tetragnathid spiders, all of which were previously unknown to science.

"You will see them," she says. "They are quite wonderful."

And I believe her.

It is dusk now and the van slows and zigs and zags along the streets of a village. This is where we will sleep, I don't say "live" because that is hardly true; we will sleep here and we will, occasionally, eat. The village is called Makawao, which my guidebook tells me means "edge of the forest."

"Waikamoi once came down this far, then?" I ask.

"I suppose it did. It doesn't now," Rosie says.

Then she laughs.

"What Makawao really means, a Hawaiian told me this, is 'the edge of something that you don't want to go into.' " She laughs again, softly. "I don't know why. I suppose there were gods, or something, there."

The van stops in a driveway and very soon I'm asleep on a blanket on the floor, but not for long.

We get up at three A.M. but Rosie says that we are not going to Waikamoi now. We will go later.

"You cannot come to Maui without seeing the sun rise from Haleakala," she says, and she is determined.

We get in the van again and grind up the long road that coils along the flanks of the mountain. It isn't in character for Rosie to spend any waking time away from work — in all her months here she has never once been to a beach — so I wonder why we are doing this. We see nothing in the dark except the curve of ground against the lighter field of stars, and as we go up it gets cold and I put on the lumberman's jacket she has given me.

I don't have to drive, I don't have to talk; again my mind slides off the track into a bumpiness of fallow and unconsecrated ground. The darkness is a comfort, as though not having to see were a ticket to freedom; it's a ticket I recognize and hold in my hand now like a gift.

The way I think of it is this: the vast unconscious universe is netted together, something like a computer's circuitry, and you can plug in any software you choose: going-to-see-spiders software, driving-up-a-mountain software. You run the program. You do what you have to do. The conscious screen reads out a wealth of fact and detail. But I have this suspicion — and it's more than a suspicion, it borders on surety — that there's some solid-state programming in there that's up and running all the time. It doesn't make you do anything in particular, but it changes the way you do everything. Call it God if you're religious minded, or a system of archetypes if you're comfortable with the Jungian slant; I try not to call it anything, it seems not to want to be named. If you try to fix it in your sights it slides away. Then suddenly it's there; it hits you like the clapper in a bell, and you resound.

Now we're going to the home of the sun god, no minor deity; a mountaintop where people have resounded often.

When we come to the end of the road we park and get out and climb higher over cinders that grate underfoot like charcoal. Clumps of silverswords — rosettes of plants furred against alpine light and wind — glow like fading stars, like flames of moonlight.

Now there is only air to look into. I think of the hoary bats, which have the distinction of being Hawaii's only native mammal and the double distinction of being rare; I think of them flying in this night air, sifting it with sound waves, swooping on moths. Now they will be circling back toward their rock crevices and tree roosts to sleep in peace. I think of the dark-rumped petrel, which come in from the sea in March to breed on these high slopes, laying their eggs in burrows and coming in from the sea only under cover of night to feed their young. The old stories tell of the night cacophony of their rookeries — parents calling to young and the young to their parents — and the stench of bird dung and fish, the noise and whirl of a dark fecundity; there were tens of thousands of these petrel in the slopes here, once. Now the rookeries are scant and frail, the eggs robbed by rats and cats and mongoose that have jumped ship into fertile ground, the young so confused by the false stars of streetlamps that on their fledgling flight they sometimes miss the sea and come down instead into highways, suburbs, parking lots.

I think of these things while light bleeds into the world from the air; the light is cool and portentous, like a birth; a sliding into being. Now shards of lava appear like broken pots and black cinder fields appear and then black cliffs, and below us less a crater than a world — a black valley gaping and falling away, a valley the size of Manhattan and more like moon- than land-scape. A kinder range of gray cloud marches across the eastern sky, with the clouds' bellies flat as if they were sailing ships half submerged in air. The sky beneath them is suddenly stained orange with a wash as opaque as paint.

I see that day comes here, too, like a shutter falling. The atmosphere thickens with light, first molten with flame and then burnt blue. So the night is extinguished, and the air congeals into a tarpaulin that spreads over us altogether.

The sun rises like an afterthought. There is no haze to turn it into an orange ball, and it rises like fire and one cannot look it in the face.

The Hawaiian sun god Maui did more than look it in the face. The night world was not his world; in fact, he didn't like it much at all — hardly a new feeling, almost human, really — so it seems meet and right that we have come here now to give a nod to what he was about. This was (or is) his house — Haleakala means "house of the sun" — and Maui was intent, as all gods seem to be, on increasing his hegemony. He wanted the days to be longer. He plotted, he planned. At last he wove a rope from his sister's pubic hair, and casting this upward one dawn long ago he caught the sun between the legs just as it rose from the sea. He then made a bargain with the captive star: that it should slow its course over this island, always, so that the days would be longer here than anywhere else.

Maui is to be admired, I suppose, for his strength and his aim, and for having made such a binding fetter out of what I have

always considered to be a scanty resource. Perhaps there is more
to this story than is easily or quickly told. There usually is. I'm
impressed by the fact that the bargain was struck here, in this
place, though whether its terms are still adhered to no one can
say. In spite of Maui's machinations there is still enough darkness here for night creatures to make a living, and for people to
live in ignorance and even fear of them; or in curiosity and awe.

Rosie and I stand in silence in the cold mountain light; we
have made our visit and it is time to go. The power of the sun is
one thing, the power of its absence is another. As we go back
down the mountain and turn around its slopes to the east, I think
again of the spiders of Waikamoi; they also spin wonders out of
fragile stuff, though it isn't light they need to capture. Nor is
capture the only thing they spin for.

It isn't until dusk that the action begins. By the time the evening
comes we are high in the forest, having come up miles of narrow
trails and through a deep ravine. As the sun drops to the west
Rosie is still counting and inspecting her spiders; there are a
hundred and four of them in her study. Each spider is marked
and the leaves of the plants that they live on are marked, and she
has to look at every one of them every day to see whether they
moved house last night, or not.

Many weeks ago she found some females guarding silken
packages of eggs. She sewed the mother spiders and their egg
wallets up in cheesecloth sacks, large soft prisons, so that she
could watch what happened as the spiderlings emerged and
grew. She now knows that some spiderlings are born with color
while others are plain, but in some of these a pattern begins to
appear after six weeks, and she has seen the colors change from
black to pink to red. There are many puzzles.

"I don't see why this should happen!" I have heard her say
three or four times already today.

The happy-faces live underneath large leaves that protect
them and their webs from the nearly ceaseless rain. In daylight
the happy-faces spread their legs and flatten themselves at the
leaf margins and go still. You can tweak the leaf then and they

won't move. Looking up at them from underneath they are impossible to see; their translucent bodies filter light through as the body of the leaf does, what markings they have look brown or black, like natural leaf holes or imperfections. Unless you turn the leaf completely over the spiders are invisible. One explanation for their varied and peculiar pixie markings is that an endlessly variable species makes it impossible for any spider-eating bird to come up with a "search image" that holds true. Perhaps this is it. Perhaps not.

I learned to find them, under their favored ginger leaves and *Broussaisia*. Other insects and snails were under there, too, night lives in hiding.

Now I'm on a hilltop thick with ferns and brush, dominated by one mammoth koa tree. The tree is a comfortable place to sit and wait. From here the ground plunges down; through the branches I see the island laid out, even a strip of white sand and surf, as pale and far off as thought.

The afternoon fades, gilding the treetops. In the slanting light I see that silvery strands of spider silk are everywhere, like tossed tinsel. There is the calling of distant birds; a massive sighing of leaves. The forest is a tapestry of the richest imaginable greens — the curved gray-greens of koa leaves, the glossy dark greens of ohia, the fronds of ferns; it's a green thickness woven into sun and mist. I'm waiting for it to fade, to rise and disappear like a theater curtain; to reveal the life.

I am not afraid as the sun goes down. Nothing here cares about me one way or the other. Perhaps even the native gods won't care much about a woman in a muddy yellow rainsuit sitting on a koa branch, a branch as broad and comfortable as a sofa.

A flock of red and black apapane have come to the koa too. It is bedding time, time for grand interactions. They squawk and trill, bobbing to one another; they whistle and click and shout. They have a vocabulary that changes constantly and reminds me a little of a mockingbird's, except that this cacophonous music is far less stereotyped, more varied and flirtatious and conversational; now the apapane come very close. They do not seem to care about me either. They feed in the red tuft flowers of the ohia and stick their whole faces into the yellow pea-shaped flowers of a mamane that is in full bloom. These are native Hawaiian birds of the honeycreeper tribe, much of which is now extinct or very rare, except for the amakihi and the iiwi and these. Honeycreepers live only in native rainforest, and there isn't much of that.

The tragedy is simple: a forest undefended is easily killed. Paradise is fragile. Don't ever forget this; don't ever forget this, I whisper to myself, as the dusk comes to the valley below me, a dusk as blue as smoke.

Even in the village of Makawao, right down there, there is not a single native Hawaiian bird, or insect, or plant, or tree.

Even this high place is not immune. Pigs gone wild uproot the forest, rats eat nests, young, fruits, flowers. Imported mosquitoes carry bird diseases against which the honeycreepers have no defense; their feet swell and they fall from their perches in the dark. The lava-tube caves on this island hold telltale fragments of seeds and leaves of plants that once clothed the slopes now planted in pineapple and condominiums and imported shrubberies. The caves hold the dry bones of birds — fifteen species of honeycreepers, ten species of geese, two species of crows, eight species of flightless rails, three species of flightless ibis, three species of long-legged owls, a hawk, an eagle, all of them extinct for centuries, all eaten or competed into oblivion by people and their domestic retinue.

Perhaps the old gods were right to warn us away. Now we can brush them aside like so many fairy tales. So they may one day brush us aside, into smoke and emptiness. There is balance in all things. I have faith in this.

Balance: I am getting to this. All day I've been revolving around spiders like a wheel about a hub, and spiders' lives are a balancing act in literal terms. How do they work? Their world is made of perceptions, as ours is. What's the world to them?

Spiders do have eyes, often eight of them, arranged like the running lights on a fancy truck, but aside from the shifts and shadows of movement no one is sure what they can see. Shape and form may be beyond them, color too. The webless hunting spiders have better vision than others, and this stands to reason, since they live by stalk-and-ambush tactics rather than the classic weave-and-wait. All spiders have some chemosensory receptors — the sense we rather falsely divide up into taste and smell — but its use is subtle, as it is with Banza or ants or flies. Some have infrared sensing equipment and can "see" heat like a pit viper does, but this talent is also best developed in the jumping spiders, which make do without webs. What web-spinners feel acutely, what they use the most, is vibration.

Their webs are like systems of telegraph lines through which they receive news of the world, and in which they hang suspended from eight legs spread wide, legs like articulated glass wire.

Something bumps a strand . . . is it a raindrop? A bird beak? Is it an insect too small to bother with or too large to deal with? The vibrations say. The wingbeats of favored prey set up a thrumming; it is time to pounce, wrap, bite. A predatory crab spider blunders into a trip-wire, and it is time to make oneself scarce. All this the silk transmits, and more.

The tinselings in the treetops are networks of knowing. The world hums through them with every breath of wind, with the arcing pulse of every caterpillar's footstep, the bump of every moth. Like the fabric of our minds they are bridges home and traceries of journeys ventured. If one could sense them as spiders do, they would be as personal as fingerprints.

I have come down from the koa; it is nearly dark, and we are sitting on a log by the path near a large colony of happy-faces. The colony features two large cheesecloth bags that contain mother spiders and their young; we have vials in our hands and are busy catching tiny flies to feed these captive families. The flies land on my muddy rainsuit and I put the vial over them, glad to be helpful.

Rosie tells me that once the spiderlings are hatched the mother happy-face has to care for her twenty-odd young for more than six weeks, trapping food for them, letting them crawl all over her. When her eggs were in their silk wallet, the mother spider crouched over them like a tigress and would take on all comers, dashing out to chomp any sneakers and would-be robbers; unguarded eggs are always eaten by something, Rosie says. The mother spider guards them with her life. She's fierce, nasty, on edge. Once the babies have emerged, things change. Her aggressive warrior ways are gone with the spiderlings' hatching and she is calm, generous, central. She hunts for them and brings prey for them to eat. Then, after six weeks or so, she dies.

"The mother spider rarely eats anything herself," Rosie says. "When she dies, and I think she starves herself to death, really, when she dies her babies . . ." Rosie looks up; "her babies don't do well at all."

Her voice is soft in the gathering dark, her head bowed.

"They just sit there in a huddle," she continues, "or they travel around from leaf to leaf in a disorderly way."

There is a silence while we think of this; we go on catching flies, filling our vials with baby spider food.

Then Rosie tells me more. In the weeks that follow the mother spider's death the family of spiderlings gradually comes apart, individuals going it alone, pairs splitting off, the clump of siblings shrinking. Mortality is high among them then, it's a tough time; as weaning is among mammals, fledging among birds; implying that the end of childhood is the most difficult transition to negotiate in all nature, that it is this which selects survivors as much as anything else.

When I look up, the night is here. The sky is a gray wash pricked with stars against which the leaves and branches of the forest are very dark. We get up then and turn our headlamps on and look at the happy-faces. They have all come down from their leaf margins and some are building more web, crossing and turning like shuttles; others have caught prey and are wrapping it busily, flinging silk around their minuscule victims; others are already feeding, hunkered over bundles. All of them are suspended in the network of web strands below the leaf roofs. Each spider tightens or splays when it feels our breath or the wobble of a branch we've touched. I have a sense of miracle; the pixie trinkets have come alive.

We count the spiders who already have prey. Six out of a colony of thirty-odd already do; it happens fast. When we've counted them, work is finished. Rosie's notebooks, which are spotted with mold and wet and have the worn softness of cloth, are tucked in her pack. My red notebook is tucked in mine. Our packs are hoisted and we stand; the margins of sight have narrowed to two yellow beams that sprout from our foreheads like cyclopean eyes.

The boles and branches of the forest trees have been sucked of color and are black. The tapestry is gone; night has given the trees the gift of attitude. Each bole has the stance of a torso, the branches are indubitable arms, the twigs are fingers or fists; they dance like so many Sivas, striking the many-armed poses of giving and taking away.

We turn back down the trail. The rainforest soil feels like mayonnaise under our feet. I didn't feel that before; now I do. I feel the variable roughness of bark under my fingers. As we descend through the trees I reach out for a handhold and find myself gripping a thing that is hairy and round and as hard as wood. When I shine my headlamp down, I see that it's a tree fern's emergent leaf, the size and shape of a bass fiddle's furled terminus, clothed in hair as red as an orangutan's.

Suddenly there are spiders everywhere, at every turn. New-spun threads shine; the wavings of feet play in the air from almost every branch, like harpists' fingers.

I stop, suddenly, having seen a great pale sheet web the size of a pillowcase.

"Rosie," I whisper, "what's this?"

"It's the web of *Labulla*," she says. "Do you see where it narrows, to a sort of funnel that goes under that bit of bark?"

I nod.

"That's her bedroom."

"Can I touch the web?"

"Not unless you want to risk Mrs. Labulla's wrath!"

We laugh.

"Watch!" she says.

The web is a level fabric, tightly woven, like crazed enamel. Over this sheet hangs a loose rigging of chaotic strands. Now there's a shift as if a shadow had jumped; Rosie grips my elbow and I see it, a spider with a leg spread the size of my wrist. Her body is coal-dark and furred with coppery hairs. She hangs underneath her web with her eight feet splayed, dimpling the woven surface.

We shine our headlamps on the web. She doesn't move. Moths are attracted to our lights and flutter; one by one they trip in the rigging and fall to the sheet, which is spread like a net, soft as a mattress. Mrs. Labulla rushes in, she bites, she waits as the moth struggles and subsides; she drags the moth down through the fabric, tearing a hole. She goes off to wrap and store her prey. Then she hustles back for more. I can see now where the sheet web has been torn and patched many times.

A little farther on there is a Broussaisia shrub and I look and sure enough there are happy-faces there, suspended under their leaves like acrobats. One is busily building a web; a formless web, a few strands loosely linked. She has a rich red "smile" on her abdomen that's hard to miss, like a daub of lipstick. Rosie's eyes meet mine in silence; it's like being at a play when a good one-liner comes along and you look at your companions to see if they're laughing too, and they are.

Male and female spiders are easy enough to tell apart. The first pair of appendages in spiders are called the pedipalps, and in the male the tips of these are swollen, budlike. The females' are plain, like simple antennae.

Rosie tells me this: that after the second-to-last molt, a female happy-face's web changes. A scent is deposited in the strands. A male happy-face will come across a scented strand during one of his nocturnal wanderings and he'll stop and take stock; then he will turn and follow the strand to its source. When he comes to the edge of the female's leaf, he will move in from the margin with slow fluid movements that are completely unlike his usual motion. When she senses him there, she orients herself toward him as if he were prey. He feels this and halts; having gotten her attention, he begins his courtship then in earnest, plucking the web strands and bobbing his body, transmitting a kind of dance-of-bounces song-of-thrums.

The dance — this is the strange thing — is different from island to island, from forest to forest, from one part of one forest to the next. There are local variations, accents, syncopations. A male can bob and pluck only in his own local rhythm, a female will respond only to her own local rhythm.

She doesn't always respond positively, even then. He bobs and plucks and advances, slowly, and sometimes she runs away, or she attacks, and if she attacks he may drop to the ground or just retreat; perhaps he will try again, perhaps not. It's like any courtship.

She may accept his advances by standing still, by doing nothing. She may encourage him by waving her front legs, sending taps of eagerness along the web strands between them. She may let go and hang by only her hind legs, her head dangling into the air, in the mating posture of her kind.

After his last and final molt a male's budlike pedipalps look as if they had burst open; only then is he ready to charge them with sperm, and mate. When he comes up to the female, he touches her with these, and if she hasn't already assumed her upright-upside-down mating position she does then; he finds her twin genital slits and inserts his pedipalps, putting sperm into her body.

Once they have mated he touches her, softly, and she pulls herself upward to stand fully in her web again. They will stay there, sharing the underside of one leaf, for half a week or more. During those nights they will mate many times.

When the female goes into her final and nuptial molt, the male attends her, helping her out of her last brittle skin. That night he will go away. He will die soon afterward.

She will lay her eggs in a rounded silk package and huddle over them, aggressive and defensive; a demon of protectiveness. If the leaf she is on falls to the ground she will chew the egg pack free and haul it up to a new sheltering leaf. When the spiderlings are ready to hatch, she will bite the package open to let them escape.

With the hatching of her young the mother spider's behavior changes utterly. This is where we began, with the mother's shift to centrality; we have come full circle in the dark, through a world stringed and keyed and tuned as subtly as a harp, on which countless variations are possible.

We go on. We seem to drift through the forest. It seems a shorter distance than it did in the day; we shoulder around tree boles and under deadfalls and dip our heads for branches, holding the trees hard as we climb or descend. When we come to the gulch, it yawns below us as dark and empty as the Void. There is a sound of running water but no sense of bottom, no sight of the far bank. We turn and lower ourselves blindly off the brink. There is nothing to hold onto except for threadlike fern roots; the trick is to distribute weight evenly between hands and feet and to move just one at a time; too much weight in any one place and we'll go down in a belly slide and bounce. We sink by inches, kicking our toes into the soil that has the texture of chocolate icing, our faces buried in the green smell of crushed ferns. At last my foot touches stone and I stand, we stand; there are stones and the glint of water; stars, the pallor of sky. We climb the other bank, hauling ourselves up through the same near nothing and then the forest is over us again; high and convoluted, full of dark trees whose branches writhe and dance, and we go on.

Here there are other spiders. Here are the Tetragnathids Rosie had promised that I would see. We come upon them one after the other; they are right beside the trail, sometimes over the trail.

The Tetragnathids of the mainlands are, Rosie says, "boring" spiders, brown and undistinguished, except for the fact that they spin some of the largest and most perfect orb webs in the spider world. Not long ago as these things go, some of these boring sorts parachuted in here from the Far East and set up shop; finding the rainforest filled with nocturnal opportunities, they began to diverge, to speciate, to set up different shops. As Darwin's Galapagos finches have different bills, the Waikamoi Tetragnathids have different methods of catching prey. This is evolution at its illustrative best. Rosie's voice is low, she's excited, this is wonderful.

Rosie has given the spiders names. They have no scientific names yet. They appear in no textbooks. What is known about them is known to Rosie and no one else.

"How could anyone miss them?" I whisper.

Our eyes meet again and we begin to laugh; the answer's too easy. It's just that no one comes in here at night.

The spiders are here, though, and they are wonderful. Here is the Elongate Forest spider; she has very long legs, very long, and she's spinning an orb web between two branches, taking two steps and swinging her body and dipping her abdomen and on again, her netted spiral taking perfect form.

Here is a Green Spiny Legs suspended among leaves from a single strand. She is bright emerald green with a gold mosaic on her abdomen. She hangs and waves her long legs like a swimmer doing the butterfly, stereotypical wavings, a formal stroke. It is easy enough to see (once Rosie tells me to look) that the bristles on her legs have lengthened and strengthened into spines. Rosie says it's these spines that the spider uses to trap its prey. The Spiny Legs will touch something in mid-wave and will clasp it, impale it on herself, like a living iron maiden. She does not use her silk at all except to hang from or to trace her way.

Then there's a Humpback Spiny Legs, with a bumpy abdomen mottled in golds and coppers; he also has spines that he uses for the catch, but he hunts on branches rather than leaves.

It's easy enough to see that these three bear a resemblance to one another, and to what must have been their ancestral form. Rosie has found four other species of Tetragnathids in Waikamoi, too, almost without looking, though none of these seven has been described — "known to science" — before.

We are high in the forest again well before dawn. It's chilly, but the rain is holding off and a nearly full moon is in the sky; the moon seems made of mother-of-pearl and pewter; it wears a halo of mist, like a veil. Its cold light makes the shadows darker than they were. One can stare at the moon, this is the thing: the sun will not bear scrutiny but this cool circle will.

The Tetragnathids are busy still. Mrs. Labulla tramples her web, still unsated. The happy-faces hang suspended, balanced, ready for anything under their leaves.

Around the moon the sky slowly lightens to a silken mauve. As the sun brings color into this perfect dark, I remember that this is the third dawn in a row that I will see, and already it seems too final, like a sunset used to: the end of the show.

The shy tweet of a roused bird, the crepitations of leaves in a dawn wind; the treetops are stained with green. The tapestry has been hung and it's over. When I look for them, the spiders have gone.

Once it is fully light, I go back to the koa again to watch the forest be and to fill in my notes. The koa holds me in a natural curve, the forest has been playing its game with me, firing the neutrinos of spider lives into mine.

I realize now that you can be as hard a scientist and as efficient a technician as you like, but the fact is that action observed makes us feel. Scientists try hard to and do succeed in filtering emotion from their results, but it remains in them. Acts elicit feelings. It makes no difference whether the acts are carried out by people on a stage or by spiders under a Broussaisia leaf after nightfall; something happens to us, having borne witness.

This happens to everyone. People who follow soap operas follow them as if they were real, they write letters to the stars warning them of treachery or advising them about their love lives, though both treachery and love are bloodless figments created by screenwriters and dispensed by media corporations; to the fans of the soaps the essential falsity of what they've seen matters not a whit. The difference with spiders is — the difference? They might as well be from another planet for all they care! But they aren't, they live here. That's what they do: live. There is no act here of passion or savagery that's put on for our benefit. Spiders are not conscious of our voyeurism anyway. Their meticulous constructs are made to profit no one but themselves.

I am not sure anymore that this is true. Last night in Makawao we ate avocados we'd found under a tree and drank wine and our tongues were loosened, not so much by wine as by what we had seen, and we talked about the men we loved, the children we wanted, the people we grieved for. What can I say?

Night in Waikamoi translates to one commandment: live. Live with every molecule, in every moment of grief and work and rest. I feel that I've been granted permission to be as I am, to indulge in the deepest and best of desires. So, in the beautiful forest, I fall asleep in the tree.

CONNECTICUT   April

As the night deepened, the wire mesh became invisible, a passage opening into a lighter blackness, and before my eyes the stars lit. They were the windows of a faraway city. They were the points of nails pressing through tar paper. They were a field of glow-bugs, motionless in time.

—Michael Dorris, *A Yellow Raft in Blue Water*

Motionless in Time

When you've grown up in a place, much of what is there is interpreted forever by the myth-mind of a child. What you have is an intimacy that is mutual, always.

This is what I have with these hills. Though I've never loved them as well as I should; something about them was always too proscribed, too deceptively calm, but there you are. They have been mine. Even decades later I still know things here the way I knew them when my own borders spread beyond the flesh. I breathe the air the way I breathed the scent of my mother's skin. I know the sounds. The seasons here beat by like my own heart.

It's ten o'clock on an April evening and I'm back, now, on the land where I grew up. This is the house I lived in, the house where my mother still lives. Nothing much has changed. Around me are the same woodlands and meadows and solid houses, lawns and gardens, ponds and brooks, the familiar soft ridges of civilized land that guide their running water out toward the sea. I have come to find what I never looked for; to see what I never knew was here.

The veil of darkness fell in my childhood as it does now, and in those days bedtime came with it; a perimeter against which I battled valiantly and long but which I never managed to press back too far. My father, coming in to say good night, would open the window. In the morning, when he came to wake me, he would close the window again. Between opening and closing the night came in: the soft rattle of the woodfrogs' spring chorus, the belling of peeper frogs, the bullfrogs' burps; and the rasps and rattlings of katydids and cone-heads and crickets; the barks of foxes, too, the coughings of deer, the yappy yodelings of screech owls; and sometimes, in winter, the deep hoo-hooting of a great horned.

What I never saw was immense, fantastic, as full as the world I did see.

I've come back to this piece of country now like three persons uncomfortable together, waiting to be healed by the sorcery of laughter or the joined labor of discovery; hopeful of everything or nothing at all. One of these persons is always gritting its teeth against the others, and all three are stubborn; by this I recognize the single sustaining thrust of what I was, what I am, and what I will be.

I have taken over a shed in my mother's garden to serve as a field laboratory. A skunk or a coon came in here in the winter and smashed flowerpots, tipped the faded tins of fertilizer off their ledges, ripped and strewed a bag of lime. By the light of a single bulb overhead I survey the wreckage and reach for the fallen broom. I notice that the broom handle has been gnawed, for the salt of the sweat that has darkened half its span.

When the mess has been swept and bagged in thick plastic, I bring in my own paraphernalia, leaving the door open to let out the dust and let in the night air and the singing of the peepers. Their ringing song, like sleighbells jigged forever just at the far line of trees, is a sound that seems woven into the stuff of my lungs, the line of my jaw.

I put empty jars on the shelves, reminding myself to bring a hammer and nail to make holes in the tops. (You always make the holes from the inside of the lid; that way if anyone gets scratched by the torn metal it will be you and not your captive. I learned this long ago. Who taught it to me?) I bring in a pile of cloth sacks for small mammals and snakes, a stack of small metal Sherman traps for catching the small mammals, and a box of the empty frozen-OJ cans that will serve as pit traps for insects. Then I lug in the four large Havaharts for catching skunks, possums, coons; though I already know that I won't use these, since this is these animals' baby-raising time of year, and I can't be heartless; these animals leave more than enough evidence of themselves everywhere anyhow. I find a nail for the clipboard, with its sheaf of empty data sheets and its pencil attached by a string. I bring in the bag of cracked corn, a jar of peanut butter, a cardboard canister of oatmeal, a plastic half-gallon jug of fermented goop made of bananas, brown sugar, and beer: the baits. I'm after everything, anything that moves here after dark. I fill a tin with pens and pencils. I hang up binoculars, tape recorder, rain gear. I put my rubber boots and hip-waders in the corner.

I look around. It looks like the kind of mess I always used to make: tacky and make-do. Serviceable, though.

Flashlights next: the big one and the medium one and the tiny one, and a sheet of red gel to cover the lenses. Night animals do not see red very well, if at all; if you shine red light at them you can watch them and they'll never know. What else? A thermos, a chocolate bar, a box of crackers.

At last I unroll the aerial survey map and nail it to one wall. On the outside of this same wall my mother once tacked a large poster of a tailless donkey, and ten small children were issued tails with thumbtacks in them; one by one we were blindfolded, spun on the flagstones, and loosed across the grass. That was my fifth birthday party. I am as dizzied and blinded now, as unsure of my bearings, and as determined in spite of it all to win, as I was then.

Down to Earth

Ten minutes ago I found one of the Sherman traps sprung. I lifted and shook it and there was an inhabiting shift of weight; when I looked in I saw a dark panting huddle, as large as a child's fist, shivering there in the beam of my flashlight.

I emptied the trap into my white catch bag with a gentle

shake. Now, nervously, slowly, I open the bag and look. Un-rolled and struggling in the cloth the animal is all of three inches long, a torpedo shape with a brief tail, an inch of stubby un-tapered gray. Its charcoal-gray fur is finer than any velvet, as fine as fraying silk tassel. The whole is subtended by four very pink feet on apparently nonexistent legs. At the front of it all is a pale nose that wiggles and whiffles constantly, by far the most mobile part of the creature, a fine almost piggy snout.

It's a *Blarina,* a short-tailed shrew. At least I think it is, I'm pretty sure; and Blarina is no rarity. It's not uncommon any-where east of the Mississippi and south to the Gulf Coast and north to James Bay; but it isn't often seen for all of that. Except by owls. Except by foxes, cats.

I look at him as if I'd fished up a miracle; though the miracle is only that he is in my bag, unharmed, as curious about all this as I am myself. In the normal way of things his eyes and ears are invisible in his fur, but this one has gotten damp in the trap (dampness being a hazard of traps) and his head fur is matted so that I can see what eyes and ears he does have, which isn't much. His ears are convoluted openings the shape of a quarter moon. His eyes are the size of pinheads and very black. I can see him rolling them up to look at me, though vision can't be his strong point; what he sees can be no more than light and dark and shades of gray. At the best of times he sees no more than I can see in this most moonless of midnights. He follows his nose.

When I spill him from the bag he follows his nose to the earth. In less than a second he drills like an auger through a sheaf of dry grass and is gone.

I prop my light on one knee and my pad on the other and write: *4/18, 11:57 P.M. Blarina brevicauda.* I check it in the field guide, and yes, that's it, I had it right the first time; I remember this much anyway from a field study course taken more than a de-cade ago, though most of the rest of that year — the calculus, the Elizabethan history — spilled from the interstices of my brain long ago. I have not forgotten Blarina.

My hands are shaking and my heartbeat is as loud as a drumroll fading off, because it works. My God, it works! I have hoisted something I have never seen ever, here, from the deep black hat of the night world. I go on writing, glad of the excuse for something tangible to put on this first blank sheet of carefully ruled and columned paper: *trap 1A, lot #9.*

What else? Oh here, yes, microhabitat . . . *dry rough meadow, on runway, near mown lawn and vegetable garden.* That's enough.

No sense, none, in rambling on.

The clouded sky sheds a diffuse gray light; the field is as wet and pale as if the world had just been born. I feel as though I had the meadow to myself, but this is the worst kind of hubris, especially since I have just proved myself wrong about the aloneness; but the feeling persists. Blarina wants to, has to, should be hidden.

Hidden: this is nothing new, by now, but it's always new. Most things are hidden. Blarinas own this place as much as anyone does; it's theirs, but the illusion of their nonexistence can persist forever, if you want. I don't; I suppose I take pleasure in poking holes in illusory things. I enjoy coming face-to-face with other lives, but it's always a surprise when something looks back and stares me down. I should be used to this: that the fabric of the world weaves itself constantly under my feet while it seems to do nothing at all, that this place, too, is like one of those puzzles in the funny pages: find sixteen horseshoes hidden in this picture. The picture is of an apple tree with a little girl swinging, the swing suspended from the tree's limb; I can only find fifteen. There's one in the girl's sock, that's easy. There are two, no . . . *three* in the grass!

I'm kneeling in the wet grass. I shake the trap out to get rid of the mess in there of shrew dung and loose oats. I notice, not for the last time, that Blarina smells bad: musky, strong. Later I will know from the moment I pick up a trap whether it's Blarina that I've caught, and a nose-to-the-ground sniff will be enough to tell me if Blarina has come this way.

I have twenty traps to check tonight, but this one has to be reset and rebaited first, so I do that. The traps are stainless steel and are two inches square in cross section and a foot or so long. To set the trap you fold one end door down and it catches on a bit of metal. The bit of metal can be pushed or pulled with a forefinger and adjusted to make the trap as sensitive as you like. This door springs up and shut when something touches a platform at the back of the box, and behind the platform is a space to put bait.

After the trap is cleaned and set, I sprinkle in rolled oats. Soon enough I'll figure out that there are problems with rolled oats: they wander around in there and gum the springs, and get musty and unappetizing after a few nights on the damp ground. Soon I'll discover better stuff: rolled oats mixed fifty-fifty with peanut butter and a little water just to the texture of putty. A ball of this cookie-ish mix can be tossed in the trap with ease, and stays there neatly and doesn't mold, and if the top comes off the bait jar in my pack it won't spill into everything and make a mess.

Tinker and refine, learn; this is what makes life good, this is what makes living.

So: if all this fuss and fiddle of traps and bags and baits is for anything at all, it's because with these I can get to the midpoint, the center line of the night world here, the node to which so many other things tie up.

Most of the small rodents here (or anywhere, for that matter) are nocturnal, and the shrews are active both day and night, so say the books; and I'll catch what I'm after if I set the traps well and in the right places. I want to know what small mammals live here, and where they live, and how they're busy in the dark.

The small rodents eat grasses, fruits, mushrooms, seeds. The shrews are related to moles, they're all insectivores, insect eaters, for the most part, though they don't stop there. Slugs are like puddings to them, snails are like biscuits; they'll eat an earthworm as if it were a long sausage, holding the wriggling end down with one pink paw. I've seen them do it. Earthworms aren't to be despised: even foxes feed on them from time to time.

Earthworms are nocturnal, too, at least as far as their above-ground business goes. Anyway, what small mammals feed on is basic stuff: push apart the grass in the meadow and the duff of the forest, right to the earth, and that's their world.

They live there, they eat what's there. Deermice will forage in trees from time to time, and boreal red-backed voles will, too, hiking up after cones, fruits, birch catkins, fungi. Blarina is an oddity in this small-mammal brotherhood: she will take big game. Blarina has the dubious honor of being one of the only truly venomous mammals, and the venom is in her saliva, as it is in the saliva of some snakes. The toxin is for stunning and killing prey, and is for defense too; perhaps that's why Blarinas smell so strong, to warn things off. Their bite is as painful as the jab of a dozen wasps. They'll eat young mice when they can get them.

Small mammals are a food source for owls, foxes, weasels, coyotes; night hunters all. They're attached to the world both ways: eat and be eaten. They're strings in the night guitar, attached at both ends, there to be plucked. What's the music? One note, one Blarina, that's all I've plucked; there will be more.

There is another reason, too, for the traps. My trap checking at dawn, dusk, and midnight gives me the excuse to tramp, to cross and recross the two hundred acres of ground that are, here, all mine from dusk to dawn; an empire with which I am well pleased.

At dawn the hegemony of this ground reverts to its legal owners. There are twenty landowners here, and seventeen houses. I have mailed questionnaires to all of them and have received them back, duly filled out, so I have a head start now on what lives here in the dark, inside and out. I know now that forty-nine people live in the houses; that after dark the people are indoors, and that most of their cars are confined to garages. Thirteen dogs, five cats, two tankfuls of tropical fish, two parakeets, a rabbit, fourteen gerbils, six hamsters, are all in with them. The land goes back then to those who have no deeded claim.

I am proud to be one of these. Along with the filled questionnaires, I have twenty signed permissions in my shed, which give
me a two-week lease to do more or less as I please here in the
dark.

Most of these came easily, others did not. One woman didn't
return hers for weeks. Finally I called:

". . ."

"Who are you, really?" she said.

I told her.

"Why are you doing this?"

"It's all in the letter," I said. "The letter I sent."

"Just tell me," she said.

I told her.

There was silence.

"Hello?" I said. "Are you there? Hello?"

More silence.

"Is all this for real? You're not trying to sell me anything, are
you?" she said at last.

"No," I said. "I just want to know what animals are in your
woods at night."

"Why in the *world* would you ever want to do that?" she said.

I told her again.

"Well . . .," she said.

Why breathe? Why live behind a six-foot fence with two
trained Dobermans? (When her questionnaire came, and it did
come, these were the facts.) Why exist? I have no answers.

So. Now, this is mine. I take an unreasonable joy in moving,
unfettered and unseen, over this piece of so well divvied and
taxed and surveyed earth. The truth is this: this suburban
countryside is richer in wildlife than many a chunk of virgin
wilderness.

I am pleased with the place even in the rain. Now the rain is
falling again, not hard; a spring rain, a farmer's rain. I load my
scattered gear into my pack and pull my wool hat over my ears
and go on toward the bridge.

The rain falls across the dark mass of the woods in streaks of silver. Here by the stream the world is awash. There are puddles between the clumps of old grass, through which new dark points are just spiking upward, each grass leaf pointed at its tip and creased along its length. The rain falls; there are glares and distortions in the air as if a curtain were flapping, a silver cloth of smoke and metal. I pull up the hood of my rainsuit and walk over the bridge. Just here there is a pond but the voices of its peepers have gone in the rain, as if they had been smothered by noise, though the rain makes no noise of its own except for a patting swish; the softest of ripping sounds, like silk being torn.

I turned my light off when I finished with the trap and now I go on through the woods, guided by the memory of the path's course and by the feel of the ground underfoot.

Trodden paths have a feel. The ground is hard. Leaves have blown off them or been trodden in. If the ground softens or I kick loose leaves, I know I've strayed and can step sideways until I feel the solid ground. Deer trails have a spiky or lumpy feel; spikier when freshly used, lumpier with clods when the deer have been running fast, bumpy when the earth is dry on a rise of ground. Human-used paths are more flattened and dished. I have learned to walk in the night with my legs slightly spraddled and my knees just bent, and to touch the ground softly with my foot sole before I tread hard, so that hollows or rocks or roots or sticks will not trip me or slow the rhythm of my walk.

Now I see a lighter loom ahead and silhouettes of branches and I know I'm coming to a road. The road feels hard, unresponsive under my feet. On the other side there is a path up through undergrowth; I put up my arm to fend off branches. Then there is the sighing dark of the hilltop pines, through which the rain drips, and I move slowly in here because I know about the rolly pinecones, lethal as roller skates left on the stairs. Then I'm at the brow of the hill and the eroded path downward full of stones; I walk on the edge because the stones make noise. The rain has stopped again. It's suddenly quiet and the air feels thick. I hear the warning bark of a fox; and I know that I've been sensed here easily enough in spite of my sneaking.

Yesterday I saw the place, a hundred yards off, where the fox had killed a hen pheasant and left her tooth marks in the bone. Perhaps, tonight, she is hunting the same things that I am hunting.

There are two old fields below me. There, too, I have set traps in the small mammals' runways. The runways are where they live, or how they travel: their homes and hunting ways, their roads. The runways are round in cross section and as wide as their inhabitants, somewhere between the diameter of a garden hose and that of the cardboard tube inside a roll of paper towels. The floor half is earth, the roof is dry grass or leaves, the walls are roots and stems gnawed neatly flush. These are easy enough to find in meadows almost anywhere; here and there a dark swatch or strip of the earth floor shows. Poking with a finger, one can feel the tunnels going off, honed by scuttling passage. I've unroofed pieces of runway, pressed the edges back, and set the traps down flush with the earthen floors.

In the farthest field, where there is a damp swale among clusters of wild grape and briar brush, I find one trap sprung. There is a heavy something inside. A big fellow. I wonder who.

I crouch and unpack: flashlight, notepad, field guide, bag. The creature is into the bag with a plop and is running, leaping, climbing the bag. Feisty. I have to push him into a bag corner and grab him through the cloth by the loose neck skin before I open up. The bag corner has a little hole and before I have him grabbed a nose comes through, with whiskers. Then here he is; coppery and kicking, with big black bubble eyes and a whiskery face, a charming face, like a rabbit's.

It's a meadow vole, a male. Very much a male. These fellows mate all year, whenever they can.

Voles like damp places. They swim well. They clip little piles of grass, neatly, to store for food. They look like coppery sausages, like micro-beavers, tiny earless guinea pigs. This one will spend a day in my shed and be photographed before I let him loose again, here, tomorrow night. I scrabble for the mayonnaise jar in my pack. Mayonnaise jars are great stuff. This one is ready, packed with dry grass.

I talk to the vole while I'm busy. Lots of biologists I've worked with do talk to their subjects, so I know there is no disgrace in the thing. Once upon a time I might have been burnt as a witch for gabbling so freely and fondly with my familiars, and for hanging bunches of weeds in my shed, and for writing obscure spells on bits of paper.

I note time, trap, lot, microhabitat, and pack the jarred vole away. I hoist the pack and shrug it on.

I go on with my rounds. The other traps are open and empty.

When I'm back by the bridge and almost home, I hear something else. The sky is clearing now and I hear the trickle-chuckle of the brook; and then splashing, and low conversation. I stop in the shadows. Now I see: two mallards are in the pond. One is following the other. It's the male doing the following, I can see the white neck band. Then he chases her around and around, there is more splashing; when he stops for breath she stands on her tail and fans her wings and gives her head a shake and looks at him sideways. They drift. Then he chases her again. I can almost touch them when they whoosh by the bank.

They mate, twice, and I find myself looking away while they do that, slowly, noisily. If I walk out I'll disturb them; I don't want to disturb them. I wait in the shadows.

After half an hour they get out on the far bank and walk off, quietly, under the willows.

I know that once they're hidden they won't mind me, so I go out through the field and up the garden to the shed. I turn on the light and set the jarred vole in a safe place, and talk to him while I put the other things away.

"Why in the world *do* I want to do this?" I ask the vole.

The vole rustles in his jar.

"Why do I get so excited about you, anyway?"

The vole peers at me through the distorting glass and wiffles his nose.

"I know, I look like hell. I know I wrecked your night, but you made mine. So we're even."

"... !"
"What do you want now?" I ask.
"......"
"Want some coffee?" I ask.
Night is a true leveler of souls.

At four in the morning the sky is clear and there is frost on the grass. Three deer are feeding in the field, gray and slim and goatlike against the white ground. They stand and watch me as I pass less than fifty yards away. It's a doe and two yearling fawns, their coats shaggy. My binoculars magnify the light and I can see their white bib-and-tucker necks and chins, and they are watching me, only curiously, like horses would watch someone passing their pasture. The doe stamps, and lifts and wags her tail in a doubtful way; one of the yearlings bounds a few paces and stands; the doe takes three steps toward me and puts her nose to the ground, to smell me out. Before she's sure of anything much I'm over the bridge and gone.

This morning there's a masked shrew in the far field trap where I caught the vole last night. They must share the runways. Why not? They don't compete. The vole is a vegetarian; the shrew eats meat. I shine my light on the runway and I can see the little snails there, crawling on the warmer earth, snails the size and shape of smoked glass beads. And a slug too, and a pill bug.

The shrew is so light in the trap that at first I think there's nothing there. She weighs as much as a wad of wool, a crumb of earth, a leaf. I can't get over her. She's the size of an unshelled peanut. She weighs (according to my book) between one-fifth and one-tenth of an ounce; less than a peanut. She's the color of earth, a dusty brown. She sits in the palm of my hand completely unafraid.

She reminds me of a miniature awl; the globular body and the whiffling pointy nose. She has an excessive frill of whiskers that feel ahead of her like cobweb fingers. Her coat is like satin.

This is the tiniest animal I've ever seen. The opposite number of the blue whale, the elephant, this bears young and suckles them. This! She survives at the bottom limit of hot-blooded possibility. According to my book, she's one of the smallest mammals in the world. The pygmy shrew is smaller, but not by much; the pygmies weigh (says the book) as much as one-seventh of an ounce. The difference is almost too close to call. In order to manage, she eats more that her weight in food every twenty-four hours. She'll eat twice as much as that if she can get it. Eating is life: day and night mean nothing, neither does weather. Masked shrews live all through Alaska and east to here, to the coasts of the North Atlantic, and everywhere between: the northern plains, the Rockies, the Appalachians.

The book is open on my other knee. It tells me these things. Trappers like me have found these things out, and these are interesting things, but what matters to me is this iota in my palm. I can't get enough of her. I hardly feel her on my hand. She's taking it all in stride, exploring my fingers, my wrist, my palm; sniffing, stroking me with that waterfall of whiskers.

Then she walks off my hand and goes off down her runway. She stops to eat a snail; I hear the crunch, I see her swallow. She's a round silky ball with tiny gleaming eyes. She goes slowly, whiffling all the way, working the runway with her nose, tapping the air like a blind woman with a stick. She's gone.

It's five-thirty now and gray-light time; I stand in the field not wanting to leave. I want to shout, to wake everyone up with a war-dance yell: *Do you know what I found? Do you know what's here?* A madwoman dancing in the hoarfrost. It wouldn't do.

The song sparrows are singing and the cardinals, too. A hawk comes over, a red-tail; it goes down clumsily into the pines. Crows and jays gather there to jeer and circle and dive at its head. I sit down on my pack to fill in my notes; the world goes silvery and gray and there is no color yet, but the night has gone, the way it goes, gently. Then there are lights in the house on the hill.

It wouldn't do. The world of people has emerged, like the

shadowy emergence of a photograph coming up in a chemical bath; hard clear forms assembling on the paper.

The silky atom in the grass has gone about her business, leaving me to mine. I have to go home. Now I want to go. With the dawn light the world has become too big, I am too naked in this open place. All my claims are null and void, until the dark.

At the Fringes of Light

I have learned to imitate the soft rolling whinny of a screech owl. I can do this well enough to hold a conversation of sorts. I imagine that it goes like this:

"I'm here. Who's there?"
"I am. Who're you?"

Something is lost in translation. What these sounds convey is beyond the capacity of language to transcribe. Perhaps it belongs to music, to the music in which you close your eyes and curl your toes.

I call, I listen. For the last two nights I have heard that whinny of reply coming in the dark, losing clarity and gaining echoes as it rolls up the valley through the leafless trees.

Now the tree trunks rise around me again and their branches lace overhead in a net in which there are stars. From the river below comes the sound of the water's continual marital argument with the stones, in voices as heavy as pots crowded together or as light as laughter.

The river runs through the woods for miles, to the sea. Stone walls cross and recross the slope; the old pastures here have been reclaimed by oak and beech. From far downriver the owl has called back to me now for two nights running. They were quiet, bright, windless nights, as this one is.

He has not answered me tonight. I called to him half an hour ago, cupping my hands, letting go my best of musical purrs. I waited. Then I called again. And again.

I am fed up with owls. I sit down on a stone. Perhaps he is fed up with me.

My vocabulary anyway does not do him justice. He can make a battle-cry whinny, a trilly love call, and will sing gurgly duets with his mate in the spring. He can let go with a torrent of high-pitched growls and yelps and can scare off intruders by screaming like a ghoul. His young make begging peeps and quawks well after the time when they should have been doing their own hunting. I'm sure this is only a sampling of the screech owl's talk, that there is more, of body language if not sound; I am sure that much is said by owls that we do not hear. I do know that each screechy has its own voice and its own renditions, her versions being higher and harsher than his, though the female is larger. I know that screechies are noisiest at dusk and dawn, and that they hunt then, too. Perhaps I have come too late.

It's a quiet night, as I said, so I hear it; the faintest of clicks, like a fleck of bark picked with a fingernail.

I scan the netted stars and there he is: a small square shape where no square shape should be, not ten feet overhead, a blocky robin-sized shape with ear tufts.

He asked a question, didn't he? Last night, and the night before he said:

". . . Who're you?"

This was my translation, anyway. Now I think he's come to find the answer out. He takes his time. I don't suppose I look like anything he's seen in his woods before.

I look back at him through the night-glass binoculars that gather more light than my own eyes can. I can see his streaky breast and his feather tufts and his great yellow eyes. I can't see the yellow but I know his eyes are yellow; they are wonderful eyes, big, quiet, set forward and together, the way ours are. All owls have binocular vision, so their depth perception is at least as fine as ours. I know that they have huge concentrations of rods in their retinas, all the better to see contrast, though they see little color. Their eyes are ten times as light-sensitive as human eyes but are shortsighted in comparison, which explains their ground-hugging hunting flight. Owls have evolved the largest eyes that can possibly fit in their skull. Some people have argued that there is little room left for brains, but that's beside the point, and unkind, and untrue; most owls have at least the same brain biomass as a crow, and crows are bright birds. Owls' eyes do take up a lot of room in there; their eyes are so large that there's no room left for eye muscles. Instead of turning their eyes they have to turn their heads. They can turn their heads 180 degrees, back to front, when they like.

They use their ears for hunting, along with their eyes. Their ears are very large and are hidden in frills of feathers on the sides of their heads. The tufts on top have nothing to do with ears. In many owls the real ears are offset in the skull, one higher than the other, which gives them "binocular" hearing so accurate that some can catch their prey on the basis of scuttle-noise alone. A mouse's scuttle-noise is not loud. Neither is a cricket's.

Screechies eat both these things, and moths and beetles, and sleepy songbirds, even grouse. For their small size they are tiger-ish hunters. They come in both eastern and western species, one or another of which is common almost anywhere in the lower forty-eight; and they don't mind living near people at all, as long as there are holes to hide in and the hunting is good. My friend Julio tells me that there have been some years — some moonlit spring nights in those years — when he has called up twenty screechies in a single suburban walk. Once — and there were witnesses to this — he was followed by nine screechies at once, eight of them females uttering tremulous love calls. This year is not like that year. All over the Northeast the screechies are at a low ebb. No one knows why.

This one is not eager to leave. There he is. Perhaps the dearth of owls has left him mateless, perhaps he thinks I have pos-sibilities, if I would leave my stone and float off, like a big coy moth. . . . I'm sorry! I'm sorry if I led you on. I don't know the words for these things in your language. Look, I'll get up and walk away, see; I'm enormous and I make a lot of noise, I have to check the traps again. I promise I won't call you anymore. I promise.

The night is full of strangeness. For once there are no airplanes passing, no dogs barking, the woods are strangely quiet. I have to climb a long hill and the open fields are pale and there is nothing in my traps. I feel suddenly bound, tied down to the ground as if I were lashed; there is no wheeling or floating for me. When I get to the road there are car lights coming so I stop behind a tree. I am against the tree, hunched like an animal, disguised as lumpiness. I am suddenly frightened of being seen. What would anyone think of a woman dressed in patched combat pants and black sneakers and a sweater three sizes too big, and a watch cap pulled down, with a backpack, and binoculars dangling, at two in the morning?

The car comes and passes. It's a cop car. After it's gone I discover that adrenaline has punched my heartbeat up as if I'd seen a tiger, loose.

The night is full of strangeness.

Every dusk and midnight I spend time hunting for owls. I do some of this hunting by making crude imitations of territorial cries, the calls that claim and carry: the whinny of the screechy, the barks and dovelike double-hoos of the long-ear. Though it's past the time of year for their singing, the long-ear's calls are too lovely to pass up. Then there are the gruntier whoos of the barred. The barred's call sounds like "Whoo cooks for yoooouu . . . who cooks for yooouu, HAAAARRR?" and is hard to do with a straight face. I like the high woodwind flute notes of the sparrow-sized saw-whet, and the low rhythmic hoot trains of the great horned. I don't attempt the whistly screech of a barn owl (which I can't manage) or the doggy yaps of a short-ear (which doesn't live here anymore, hasn't for years), but I listen for these anyway. Some owls will answer the cries of other owls. One can't be sure what a call will call in. Great horneds will hunt screechies and saw-whets. Screechies will mob great horneds. Saw-whets are shy, talkative; they keep to deep cover. Barn owls used to be common here and I've watched them hunt, in years past, when a family of them spent three seasons in a neighbor's barn. They are big pale birds with longer and narrower wings than most owls have. They like open country.

Some owls are more nocturnal than others. The arctic owls, the great grays and snowies, hunt in daylight all summer because that's all there is, though I've seen snowies hunting at midday in the winter, too. The local great horneds will strafe ducks in full daylight, in February, when they've got young in the nest and are short on nocturnal groceries. The hawk owl of the taiga hunts in daylight — hawklike — whatever the season. The burrowing owls of the West hunt day or night, it's all the same to them.

The owls that live around here prefer to hunt in darkness, though an overcast afternoon may be dark enough if they're hungry enough. Dusk and dawn in particular and moonlit nights in general are best for their hunting, provided there is no wind. Wind interferes with listening, and owls listen. They fly low and slow on lightly loaded wings, buoyant as moths. Their feather edges are frilled to silence their flight. They angle down as quietly as a breath of air. They hoist their splayed talons nearly under their chins and hit their prey with all the force of their weight and dropping flight to drive the scimitars of the talons home. Swinging the prey beneath them, they rise and float to their hunting perches. They do this again and again. They swallow their prey whole. Once or twice a day they cough up an egglike pellet of bones, feathers, and fur. Their hunting perches can be found by watching for bark or branches whitened by droppings, or for heaps of these pellets. The pellets can be picked apart to see what it is that the owls have been eating. At one time or another I have found the yellow shafts of flicker feathers in them, and the curves of vole jawbones, the matted

Eastern Screech Owl
Otus asio

gray-white of cottontails' fur, the nut-brown wings of beetles.

Owls can be hunted in daylight, too, and I do this during my
dawn patrol as soon as there is light enough to see. I look for the
"whitewash" and the pellets, I scan the trees for likely holes. I
want to stake out an owl place — nest or hunting perch — which
can be watched at night. Holes in trees — old pecker-holes,
rotted limbs' holes — are common enough. I know that if the
tree is banged on with a stout enough stick an owl face will
sometimes fill the hole: a frilled visage, barklike as bark itself,
each round eye the center of a feathered vortex.

In my banging for owls I often think of myself as nature's
equivalent of a door-to-door peddler of unnecessaries; of cleaning brushes or encyclopedias, perhaps. I do my best to knock
soundly and with confidence, in a firm but nonthreatening manner, and am ready to greet my customers with deference; I work
my territory with bird book tucked ready under my arm, the
pages of owl samples marked with a stem of grass.

I vary the times, intensities, and rhythms of my summons,
but have no success at all, though this ploy has yielded customers aplenty in other places and at other times. I begin to wonder
if all the owls in the place have grown as wary of strangers at
their doors as their human neighbors have, or whether they —
along with the people here — have simply become immune to
noise.

I often wonder about the noise. The daytime is bad enough but
the night seems worse, to me. Even without the snoring of
diurnal cars this place is almost always noisier in the dark than
wild places are. Airplanes can be annoying beyond belief, their
whiny drones come and go and come and go, jamming my
tremulous calls and the owls' more tremulous replies, smother-
ing the fainter rustlings of small game. Dogs bark; especially on
moonlit nights there is hardly a moment when a dog isn't
sounding off somewhere, and the sound carries. Two ridges

over, someone drives home late and the tires whine, stopping me in my tracks: until the car has gone beyond hearing range I can hear nothing else. By this I know that I have been listening hard, that the oblivious party-goer is depriving me, at least, of an orienting sense. On still nights I can even hear the hissing grumble of semis on the highway, though the highway is ten miles away.

I wonder, too, about the light. The headlights of cars flashing by wreck my own night vision, they're literally blinding, and I wonder what effect they have on other things. I know that owls are struck by cars often enough. On overcast nights here the cities shed an orange glow on the clouds, so it is lighter than less citified places nearly all the time, and the farthest stars are always invisible, as if they had been curtained off. I don't know what difference this makes for owls.

I don't know what difference anything makes. I note these things down. I note, too, that the noise, when it is there, makes me irritable. It doesn't bother me in daylight, though then the noise is surely worse; I notice it in the dark when I cannot see well and am listening instead.

With one thing and another my owl hunts yield not much. That single screechy is the only contact surely made, though there may have been others of which I was unaware. After that night I keep my promise to him and I call no more, though I do go on with my cnoking bouts. Once daylight has come, I listen for cacophonies of songbirds, too, especially jays, who will mob owls when they find them roosting in an open place. What mobbing I do hear turns out to be about hawks, of which there are plenty, and never owls.

I reach no conclusion other than the fact that my owl-hunting ways are not good enough. I heard barreds this last winter and great horneds, too, so I'm sure they are here. I'm glad they are here. We share our darkness and our hiding and our ground, if nothing else.

*　　*　　*

There are too many things of which I am unaware.

Out of the corners of one's eyes are miracles; at the fringes of sensibility is delight. I don't know where this phrase came from but it's in my head, graven, as if tiled to a cerebral wall, and I repeat and repeat it as I walk and walk here in the dark. It whirs like an *Ave Maria,* a kind of penitence.

I don't see much. No one ever does.

I can see signs, glyphs, scribblings that speak plainly enough if I care to read them. If I add them up, one to another, with diligence, a pattern emerges. I only know that nothing, however fragmentary, is without meaning. Nothing at all.

In the gray of predawn, in the zodiacal light of sunbeams bounced by atmospheric bits, the first cars snore to life and begin their beaconed city-bound prowl. I am in the midst of my predawn trap check then, and I note that song sparrows swing into gear an hour before the cars.

I wonder how many people who are leaving for the city now will see the birds. How many will see, in an hour or two, a flock of birds flying from a tree to the ground or to another tree as they drive in the morning light down Fifth Avenue past Central Park.

If I were behind the wheel of a car, I wouldn't see the birds. I would be watching the red light, the taxi bumper nigh to my port side, the weave of a cyclist. I would be watching the tall businessman with dark hair crossing in front of me, taking the curb and half turning to push the gold-framed dark of a revolving door, the briefcase corner and the heel of his shoe disappearing last. I wouldn't see the birds.

I know, in spite of the distractions of the world, that the rectangle of ground at the Big Apple's heart is a wonder.

Manhattan is the junction of the Hudson River Valley and the coastal routes, two superflyways long graven in the earth and air. Manhattan is the meeting place, the central station; and Central Park is the most habitable part of it, for birds. During the heavy travel times of spring and fall, almost half a continent's worth of birds is coming and going there.

Of the more than two hundred species of birds that one can see in Central Park, if one knows how to look, most of these come and go at night.

They have gyrocompasses in their inner ears. They can feel the tug of the planet's magnetic field. They can sense that the weather will change, or not change, as they can sense their altitude, by a barometric knowing. Some have come from Guatemala. Some will go off to Greenland.

Shorebirds, orioles, flycatchers, sparrows, wood warblers, wrens, vireos, thrushes all hurtle northward or southward in the dark and at the edges of the year. So many pass in and out of Manhattan that a hail of night migrants pelts against buildings. Brightly lit things draw them helplessly as if they were moths. The Statue of Liberty, blazing torch raised, pulls in more than seven hundred feathered fatalities every month.

These birds see better than we do. They will notice barklike moths against the bark of trees or pale seeds on a pale stone, but they do not migrate by seeing. The songbirds' navigational senses function in the dark, so that they can move across the continent safe from predators (if not obstacles) under the cover of night, but they feed in the daytime. Of all groups of animals — besides that to which we belong — birds are the most visually oriented, the most daylight dependent.

Which makes the exceptions to this rule more marvelous. Not only night migrants, but the birds that are night birds by nature, by practice, by expertise.

There are not only owls, there are others.

One of these flew down by my feet, at midnight, three nights ago. I saw it come down out of the corner of my eye and I stopped where I was in the grass.

I knew by the plump-plover silhouette that this was a woodcock, though I'd only seen two, ever, before. It flew down and along the ground, flittering, batlike. Its big black eyes were set high on its head, all the better to watch the wide sky for an owl's swoop, a fox's sliding silhouette. Its beak was as thick and long as a stick. It probed the lawn for worms and grubs just as its shorebird relatives probe the sand or estuary mud. There was a moon, and I could see these things plainly.

My mother told me, later, that my father used to go out alone on April evenings to stand just where I stood, to listen for the woodcock's "roving." True to his European past, he always called it roving, and that is what the larger heavier-bodied European woodcocks do: "rove" when they sing and "rod" as they pass overhead in their whistling courtship flight. American woodcocks, true to their diminutive nature, are properly known to "peent." My father listened for roving, which is what he heard. He never told me this. His birding was a contemplative pursuit rather than a competitive one, and he listened for woodcock, here, for more than twenty years.

Woodcocks are rare here now. They used to be common. They like damp untidy places, or what people can think of as damp or untidy; they don't mind living near people so long as there is cover by walls and fences, and small ponds bordered by uncut meadows, and wetlands that are rough but not too rough. They need soft ground to feed in. They nest on the ground and they rest on the ground. Their russet breasts, their patternings of yellows, grays, browns, and blacks, exactly match the patterns of fallen leaves. Their heavy bill looks like a tilted stick. When they're resting in daytime, they won't fly up until you nearly step on them and then they're away, on wings almost square, dodging low between the little willows.

I heard their roving once, on a spring night in Vermont. I couldn't see the birds, but I heard the whirring of the male's rise, the fluty twitterings at the peak of his flight, the whistling dashes of his descent, the low "eenp eeenp" as he landed and strutted again. This is a serenade, and is often a communal effort, with several males competing on the same peenting ground. The females wait and watch, and — presumably — listen, since the sound of it all is more strange and wonderful than anything in the air. The peenting begins at dusk and goes, sometimes, until dawn. I think it is too late for it, here, now. My mother tells me that my father used to hear it early in April. It's late April now. I listen but hear nothing. If they are here, then they are nesting, so I keep out of wetlands out of respect for them.

The woodcock isn't the only bird hidden here in the open light. Whippoorwills and nighthawks also spend their days resting on the earth or on branches, and, like the woodcock and the owls, they mimic the colors and stillness of these speckled things. Whippoorwills and nighthawks are members of the bird family known as nightjars; they have short, wide, liplike beaks made for catching insects on the wing and in the dark. I always used to think (still subliminally do) that the name "nightjar" referred to the size of their mouths. The Australasian frogmouth (a distant relative of our native 'jars) takes the prize for mouthiness; it looks like a pocketbook. Anyway, the whippoorwill is modest, almost swallow-like in looks; though at the corner of its widish mouth is a sudden frill of whiskers like a sparse and badly trimmed mustache. These whiskers act like a set of fingers or a sensitive net to feel and sift the night air.

I have heard whippoorwills singing here, just at dusk, all my life. I hear one now every evening on the wooded slope; the slope where I heard the fox bark, above the field where I caught the vole and the shrew. Slowly the nights are being peopled, inhabited, one night fitting into another like puzzle pieces: the shrews, the voles, the foxes, the whippoorwills, all there.

The whippoorwill says:

"Weeep . . . weep! Weeep . . . ooo . . . weoooo."

It's a mournful rocking sound, like an old ballad of over-the-sea-sailing, of gone-to-the-wars-my-Johnnie.

Tonight it is dark with a strong wind and I feel as if it should rain, but the wind goes on and on, rainless, across the hills. In the forest by the stream trees toss and clatter. The dying moon is in a gold-dust nimbus; where the moonlight catches the pool the water is as wrinkled as used foil. I go on over the hill and am halfway down the far slope when I stop; there are voices in the trees. One oak makes a buzzing, the angry hum of a trapped fly or a full hive, but very loud; when I look up, I see the branch of another tree wearing on the trunk of this one like the slow, sawing bow of a violin. Other trees groan, or chirp, or cry out with a high cry, or rattle bonily like stays on the mast of a ship at anchor. The trees on this slope are overgrown with vines; now in the wind their burdened wood makes unhappy sounds. I know this, but the sweet voice of reason is overwhelmed by the conviction that the trees are sentient, and malicious, and are giving tongue. The whippoorwill adds his mournful balladeering then, and I nearly turn tail and run.

Night fear is no stranger anymore. It makes my blood drum in my ears and I want badly to run, but it will go away. My antidote to panic is to sit down and wait for it to go away. I sit down here in the tangled grass.

"Weeep . . . weeep! . . . ooooweeep . . ." I hear again, and I hear the groans and cries and rattlings of the wood, and the wind sighing over the ridges. The moonlight comes and goes like waving cloth.

I think about the night birds, this odd fringelike brotherhood, less brotherhood than fringe; attenuations of one bird family or another that have gone pioneering, on their own, past the avian boundary of the dusk. It's a growing business; more birds are doing it, evidently, all the time; though evolution being what it is, this is no night-grab rush but a slow and purposeful creep. There are nocturnal curassows in Central and South America; there's the Italian occhione, a huge-eyed nocturnal shorebird; there's an owl-parrot in New Guinea. Australasia hosts some of the most bizarre of the nightjar cousinage — several potoos and, of course, frogmouths.

There are the kiwis of New Zealand, too, which look like a cross between a woodcock and a monstrous shrew. They're chicken-sized but woodcock-shaped, with a long probing bill;

they are a shrewlike gray-brown with a plenitude of shrewy whiskers and black-bead shrew eyes. They live in burrows. They defend themselves with claws. Their wings are no larger than a crooked finger. They probe in and along the ground in the dark, their nostrils are at the ends of their bills, and even their feathers are more like fur than anything.

Then there are the South American oilbirds, which look like swifts, nest deep in caves, and echolocate like bats, but with a train of eerily audible clicks.

And these: the shearwaters and petrel who spend their lives over the open sea, hunting in the dark, smelling their prey by the scent of oils in the surface waters; oils of squid and herring, oils of anchovy. When they come to land to nest, they can smell their own burrows on the mountainside, their own burrows among thousands of others.

Around here there are night herons, and owls, and woodcocks, and the little nightjars. That's all, really. That's about all there is.

I don't see any nighthawks in all the time that I walk my small domain, but they should be here. They feed at dusk and I have seen them other years, when I wasn't looking for anything at all. They are not hawks; they are more like big nocturnal swallows, larger versions of the whippoorwill. I noticed them for the first time when I was sitting on a friend's back porch a mile away and ten years ago. I saw them then out of the corner of my eye, and focused on them only because I had never seen swallows so large, or out so late, or with white patches under their wings. Those white patches are like bold identifiers on an airplane. I found out that they were nighthawks, later, when I saw a picture of them in a book.

I have seen night herons in the same way, when I wasn't looking. I saw a black-crowned night heron last spring when I was driving across a bridge through the next town — which is a city, really; the street on both sides of the bridge is lined with fast-food joints and so on. It was dusk and the night heron was standing in the river near the bridge. I came back with a flashlight two hours later and parked in the Dairy Queen parking lot and climbed over a fence and down the embankment. The night heron was there. I didn't need the flashlight. Traffic banged and rattled over the bridge. I watched the heron poke forward; I watched him spear and then shake and then swallow a frog.

I saw nighthawks last September, when they were coming through Connecticut on their way, presumably, south. I was not thinking of nighthawks then at all. Out of the corners of the eyes, miracles; at the fringes of sensibility, delight: the world showers gifts on my head when my eyes are nearly shut; when I have been drinking wine and want to go home. This is the way it happened:

It was a windy night. I had dinner with friends in that next and more urban town at a restaurant by the sea. When I came out from dinner, the streetlights in the car-park were whirling with nighthawks. There were more than a dozen of them, arcing around the lights like oversized barn swallows, with the slanted bars of white under their long brown wings. They whirled and whirled. The lights attracted insects — I could see clouds of insects and so could they. Nighthawks spend the winter in the Gulf states or as far south as Guatemala. They were tanking up, here. I had seen bats feed by streetlights, but never these. On the way home I looked at every streetlight for more, but I saw no more, though multitudes of birds were surely moving in the northeast wind, the wind that was blowing like a river; the river in which we are bound, and submerged, like stones.

Inside Story

One night I went out with Dave Norris to look for hibernating bats. I was honored that he would trust me enough to take me along. The place is known only by a select and exclusive few, and they have an unspoken pact not to give this secret away. It turns out that bats need protecting more than almost anything else, and that secrecy is the best protection.

It was dark and still and starry when we went. We drove

through the country for almost half an hour and then we parked and walked along a road. To get to the place, in an aqueduct, we would have to trespass; and that made me more nervous than I liked to admit. I was not inured to trespass, yet. We wore hip waders and carried flashlights the size of automatic weapons. I was ready to leap for the ditch if a car came.

I have a hard time understanding bats. I'm not sure why this is, but I've found out since that other people feel the same. Perhaps we have projected our own night terrors on them and have made them the scapegoat for something that is ours and that we don't exactly like.

I know that people who have had some experience with bats do not forget it. What they seem to have in common, afterward, is puzzlement; a profound knitting of the brows. I have never heard stories told with so much odd and almost religious gravity, as stories about bats.

I have a bin where I collect stories, tales, natural-history fables of all kinds. There are several in there about bats. Some of these are written on napkins or the backs of envelopes and shopping lists; some are notes made in trains and most have been pieced together out of fragments. Whatever I hear has been filtered through a human sensibility. They are stories piecemeal. Here are a few.

One friend told me about the fruit bats he saw when he went to Southeast Asia. They were very large, he said, spreading his arms. Some had a wingspan of five feet, more or less. In daytime they roosted in trees with their wings folded. They looked like bunches of leathery rags with animal faces.

"They weren't bat faces," he said. "They were *animal* faces."

These bats ate fruit at night, gulping it whole and strewing seeds through the landscape. A fig takes only half an hour to go through a fruit bat. Through it goes, to the everlasting glory of figs. Some of the fruit bats looked like little foxes — they are called flying foxes — and have cute ears and round dark eyes and russet fur. Their faces haven't had to contort to sonar principles; they neither send nor receive high beams of sound. They navigate by sight and smell. My friend said that in the daytime they would peer down at him, curiously, out of their bundles of raggy wings. They seemed to see him just fine.

"They had to twist their necks, though, to get a good view," he said. "They made me feel that *I* was upside down."

At night they'd fly right over his head, he said, whoosh and gone. He said that they reminded him of the flying monkeys from *The Wizard of Oz*. He said that when he was little he used to have bad dreams about the flying monkeys, and that the bats were very like the dreams.

Another friend told me about the bats who lived behind a shutter on his house last summer. He lives with his wife and two children in a suburban house not far from here, and commutes to his office at a prestigious law firm. He called me three times, from his office, about the bats.

Every evening at a certain time they would drop from the shutter like ragged bits of dark. They would fall straight down and then suddenly flitter up and away and around. He could see them only when they crossed the lighter sky.

He was interested in these bats. His interest was only thinly veiled by the desire to have them removed from his house. He had called the exterminators, and then had called them back twice, each time to put the whole thing off for another week. He was hoping that the bats would leave on their own. In the process of his inquiry he learned that the bats had to roost up high so that they could drop, the dusk drop-out giving them essential air speed. They could not, he said, easily take off from the ground. He was intrigued. He wondered why they had to live behind his shutter.

My mother once had a bat in her bedroom. After she got in bed and turned out the light, the bat would appear and would start flying. Hour after hour it made patterns overhead. It made whorled pentagons and bisected hexagons and other more complex and mystical geometries. It looked like a bit of animated leather that was busily obeying some wizard's symmetry. It wouldn't fly out the window. It kept her awake. For five nights it was there and for five days she hunted it, in curtains, on high closet shelves; but at daylight it had conjured itself more utterly away than any incantation could explain. On the sixth night she left all the windows open and all the lights on outdoors and she slept in my old bedroom down the hall. After that, the bat was gone.

Now I wonder what I am going to see. I haven't seen much until now, though I've seen bats flying often enough. Flying bats move too fast for the human eye to freeze-frame any shape. I have seen them flitter across the lighter sky, tacking back and forth over open ground again and again, pacing the air, like foxes in a cage. They look like nothing; like no form. I have never seen a bat close to, alive, and still. I ask Dave a lot of questions, but he has more questions than I do. No one, it turns out, knows very much about bats.

Dave Norris knows as much as anyone here. He is a naturalist by training, by nature, by dint of passion. He likes looking for bats, among other things — other things being hunting for salamander spawning ponds and trapping small mammals wherever he goes, hoping to snag a rarity: a bog lemming or a pygmy shrew. He teaches biology at the Redding Easton high school. He teaches a course in field biology there, too, and this is popular, since field work is what Dave likes most. Enthusiasm should be the only prerequisite of any teacher anywhere.

Sometime in March he goes to a cave he knows of and nabs a hibernating Keen's bat and takes it home and puts it in his refrigerator. Later, in class, it tries to fly out of the aquarium that he's put it in, but it can take off only at a shallow angle from the ground, like a heavily loaded transport. Bats don't have to drop out to get air speed, but it's easier on them if they can. This one can't, and the aquarium-runway is too short for other things, so it stays in there. It scrabbles, using its claw feet and spiky thumbs. It grins around itself, echolocating inaudibly, trying for a fix on the world. After the class has seen what it needs to see, the students take turns feeding it some mealworms. The mealworms make up for the energy that the bat has burned. A bat has only so much energy, stored as fat, to get himself through the cold and foodless winter, and this bat has used a lot of his in waking from hibernation and scrabbling around. The students learn, as the bat plucks the mealworms from their fingertips, that if they didn't do what they are doing, this one would starve before the spring.

After the bat has been fed it goes back in the refrigerator. It hangs from the top rack. In the cold and dark, suspended over newspapers, it goes back to hibernation. Dave returns it to its own hibernaculum on the following night. When he takes the bat from the cave, he marks its place with a ring of chalk, and he puts it back right there. When it rouses in April, it will be where it remembers having put itself, and it may remember a nightmare of sorts, a horror-show dream, or it may not; we'll never know.

Dave showed me his collection of bats when I first arrived at his house. They were dried and stuffed specimens in a cardboard box. Their fur was rough and their faces were muzzy, their wing membranes were as stiff as dried leaves. They didn't look great. Bats, like flowers, are too delicate to survive the specimen process with grace, but there they were. There is nothing like something to heft in the hand.

None of the specimens had been collected live. That would have been against Dave's principles. Many had been brought in by students and most of them had been collected in parking lots. It turns out that cars hit and kill bats with a fair degree of frequency, especially in the spring and fall when they're migrating through. Sonar tuned to pluck up fruit flies at a distance of half a meter and at a rate of two per second, or to scoop katydids from a blurring wildnesss of grass and leaves, is not designed to notice an oncoming Ford. A driver is not designed to notice bats. A bat in flight looks like a leaf falling or blowing by, and even the biggest ones here, the hoary bats, weigh only an ounce. Most weigh half of that, or less. Hitting a bat isn't something a driver can be aware of. Sometimes the bat stays stuck to the car. Next day, the car takes its driver off to work or to the mall, and after a while the sun dries the bat and it falls off the bumper onto the parking lot. When the bats are brought to Dave, they are still moist enough to skin and stuff and pin into a decent specimen.

Among the brownish others in the cardboard box is a single red bat, auburn furred, with a pleasant short-nosed face like a teddy bear. It spends its summer and rears its young in the northern United States and Canada, and it winters in the open in trees in the South. In summer it lives a solitary life. Unlike most bats the red bears litters of young, up to four at a time, perhaps because so many of its kind are lost to the hazards of migration. Migration is the main reason that it's found in Connecticut. Some may breed here, but for most we are en route, no more.

The hoary bat lives by much the same pattern. Dave has one of those too, its long fur pale-tipped, like a swatch of silver fox. It was also passing through.

The box also holds a little brown, a big brown, and a Keen's. These are all common here all summer, starting now, and even before now. I saw what were probably big browns back in March; I watched a pair of them hunting over a pond. There was a moon then, the sky was bright and clear, and I could see them well. It is impossible to get a real fix on a bat on the wing, one can never be sure, but these were bigger than the little brown or any of the other Myotis, bigger than the pipistrel, and big browns are known to be active in weather too cool for most bats to bother with. They hold a monopoly on early moths, gnats, stoneflies.

The big and little browns will take a summer freehold in buildings; sometimes the big browns will even hibernate in buildings. These two and the little pipistrel are the bats that people here are most likely to see. Cupolas and barns and belfries and the backsides of shutters and even open windows bear reasonable resemblances to hollows of trees or crevices in rocks. Old hollow trees have become a rarity here. A lot of rocky spots have been blasted or paved over, or have switched their allegiance to the display of alpine shrubs. Bats are nothing if not adaptable.

Within their limits; within their limits, yes.

We cut off the road now on to a track through new growth timber. I can swing an arm around these young trees as if they were friends; they run straight to the sky like systems of mountain rivers. They are only barely thickened by their swollen buds, they have no foliage now but stars. There are no hideholes in here for summering bats, or anything else.

We go over a brink, taking care in the slippishness of damp half frozen leaves. Water winks and chutters down below and soon we are in it to our knees, and our feet are on stones. Ahead of us is an opening darker than any night. We follow the river in.

This is as much a man-made place as the woods are, and more so, since the woods were woods before and this was only hill. It was reduced to hill after glaciers wore over it a time or two. Now it's been further reduced to an arc: it's pierced by an aqueduct seven to ten feet high and three-quarters of a mile long. It carries water from a reservoir to folks downstream.

Connecticut has few natural caves to speak of. In times gone by, most of the bats around here wintered somewhere else. Some went to Massachusetts, others wintered in Vermont; in both those places there are natural limestone caves. Wintering caves have to be deep enough never to freeze or the bats will freeze with them; what stays dark, moist, cool, and doesn't freeze is a bat's definition of a cave. Thanks to people's business, there are mineshafts and aqueducts in Connecticut now, and bats have discovered that these are also caves.

Just inside the entrance a phoebe has made a nest. The nest is mud wadding and we see her tail sticking out and her head. Her eyes are open but she doesn't move. We duck, and go in, and turn on our lights then, not having dared to turn them on before. Not having needed to.

Once we are in, the ceiling is high enough so that we can stand and walk easily abreast. Sometimes the water is up to our thighs and sluggish, sometimes it's rapid and shallow and one can see the patterns of its brownish rocks. The ceiling of the tunnel is broken, rough, full of sharp crevices, and drill holes, and angles. The rock is granite pinked with chunks of feldspar, flecked with quartz, glittered with the silver-black scaliness of schist. This is the bone stuff of New England: penetrated, shattered like glass. We look up into crevices and angles and holes. This is where the bats will be.

Dave says that in winter the ice goes no farther than six feet in. Then the entrance is veiled with icicles; beyond that, the groundwater drips unfrozen, and the stream is unfrozen, and the cave begins.

Most of the bats are in deeper than this. They take no chances.

It is late for bats to be hibernating here, but this has been a cold and a tardy spring. It is the end of April now and it was 42 degrees outdoors at eleven P.M., just before we came. It will freeze tonight, again. How do the bats in their deeps here, in their unchanging deeps, know this? That is one question. There have been dozens of questions already tonight. How did the bats discover the place? How did they know they could hibernate here? How do they navigate across country anyway — by what combination of instinct and learning? No one knows these things. How does one study the habits of bats? No one knows a good way to do that, either. People used to study bats here by sexing and weighing and banding them when they were in their hibernating caves. That's the only time you can find them for sure. In summer they scatter to their barns and attics, their crevices and holes, singly or in little colonies, and they raise their young then and feed then, too; but who can watch them there except when they're asleep? The problem is this: disturbing them in their winter caves turns out to be an excellent way of killing them. It is difficult to measure, weigh, sex, and tag an animal, any animal, without disturbing it. Disturbed bats burn up hibernating fuel at a horrifying rate. Some of them die. Sometimes lots of them die. So there you are.

Bats have been around for 50 million years, and there are more than nine hundred species of them around the world. The Chiropteran order, which includes all bats and only bats, boasts more species diversity than any other order of mammals except for the rodents. Some bats catch fish; big fox-faced tropical bats feed on fruit or nectar; some South American bats suck blood; all the bats that live here eat insects, and find their prey in the dark by using a kind of sonar.

Their sound beams can be shifted in intensity, frequency, and focus, and are shouted through mouth or snorted through nose, depending on the species. The snorting types' noses are like megaphones or sonic transmitters; they have leaflike protuberances and are naked and whorled like nasal ears. They look, to our eyes, exceeding strange. The shouting types have doggier noses; in spite of their size and coloring they look less like mice than like tiny bears. Whatever its source, most bats' sonar sound is too high to be audible to us, but by all reports it is very loud. High frequencies vanish rapidly with distance. This "shortsightedness" (shortsoundedness?) can be compensated for, to some degree, with loudness. High-frequency sounds do have the advantage of small wavelengths, which means that they can "resolve" a finer "picture." The sonic picture of a spider on a leaf must be exceedingly fine, drawn with a hair-fine wave, in order to be audible. Low frequencies carry farther but would make for too coarse a grain to be of use: a picture drawn with a two-inch brush does justice to mountains perhaps, and even to trees, but not to mosquitoes. Most bats do have lower voices, too, voices that we can hear; these are voices, not sonar, and are made for chattering among themselves.

Dave says that bats are healthy animals, healthier than most.

They are no more apt to be rabid than squirrels or coons and are far less likely to infect people even if they are: teeth made for holding moths are rarely strong enough to pierce human skin. In general bats are a healthy lot, perhaps because their habits of crowding up have encouraged the development of immune systems built to stand the strain.

Speaking of crowding up, there is a hint in this of the reason for bats' success, and fragility. Most bats need caves. They need them to adjust for extremes of climate, and to do what they do there in safety. They use caves to hibernate in, or to raise their young in, or both. Good caves are not everywhere. Where good caves are, the bats, of necessity, crowd up.

This one will not be crowded. This is no city of millions like the Batu Caves of Malaysia, the Niah Cave of Borneo, the Tamana Cave of Trinidad, the Carlsbad Caverns of New Mexico. This isn't Texas. More than a hundred million Mexican freetailed bats summer and raise their young in five big Texas caves, and consume more than twenty thousand tons of insects in the caves' environs every year. This tunnel makes no pretense to Texan grandiosity. This is an outpost, a village of pioneers in what was wilderness.

We go on, wading forward. Once we turn out our lights and the darkness settles on us like thicknesses of soft black cloth; my instinct is that I am being smothered. I have never been anywhere — *anywhere* — as dark as this.

Sounds echo here. The drips and ploops of water are loud; there is a metallic knocking as a wave rises in an angle of stone and sucks away.

We turn our lights on and go forward again, whispering.

There is a small checkered garter snake on a ledge above the stream. He is hibernating, too. He looks newly enameled, his yellows shine like gold, his head is as slim as a reed. He doesn't move.

Dave stops and nods, and I look up where he is looking, but I don't see it at first because it is so small. I had in my mind something dark and ragged the size of my hand; this is a rectangle of golden fur. It is a hibernating pipistrel.

Its body is no larger than the shape made by my thumb and forefinger enclosing a circle. It has little ears and a round pugged face with a slightly upturned nose; almost too cute to be really alive. Its long forearms are tight along its flanks; the arm bone shows pink, its hands with their webs of wing skin are folded on its belly like furled petals. Between its tail and thigh is another web of skin and this is tucked inward, between its legs, like an enclosing leaf. I can see its sides heaving with breath. Neatly packaged and wrapped, it hangs there from its feet, each toe clawlike, black, hooked in the granite.

Dave tells me that the pips always hibernate in solitude. One hangs here, another there.

Myotis bats crowd up. We come to a place where the roof has partly fallen in, making a heap of rubble beneath and a dome overhead; this is where lots of Myotis come, he says. In wintertime the different bats claim their own parts of the tunnel. The big browns — the hardiest — hang near the entrance. The little browns and other Myotis — the Keen's and sometimes even a few of the rare Indianas — come farther in. They come to this cavern and pack into drill holes and crevices and all you can see then is their noses and ears. The pips are more delicate, perhaps because of their solitary habits, and they go deeper still.

Now most of the hibernators have already gone. They've moved out in the order of their hardiness, the big browns first. Only three Myotis still cling to the cavern roof, too high overhead to identify for sure.

We walk in more than half a mile. Dave has made markings on the walls that tell him where we are, how far we've gone. We count eleven pipistrels, a dozen Myotis too high to identify, and three Keen's. The Myotis are all larger, darker, more wakable than the pips. One Keen's wakes as we pass and grins at us, showing a mouthful of needle teeth. He doesn't mean attack; he is shouting, shining his beam of sound, varying frequencies, sweeping us with FM, trying to discover what and where we are. Dave says these Keen's have probably been out and about already, probably several times, but have not taken up a summer residence as yet. They come back here now when the air is too cold for hunting comfort.

The Myotis are impossible to tell apart on the wing. They are hard enough to identify even up close; their visible differences are in their traguses, the small projection in front of their ears. Dave shows me. The Keen's tragus is pointed and long, not at all like the blunt one of the pips. Later, when I look in the mirror, I find I have traguses too, of a sort, those bumps that project a little into the opening of my ears. I move my fingers upward from my earlobes and I feel them there, cartilaginous as fingertips. If I compress them with my hands and get them out of the way, I find that I can no longer tell what direction any sound is coming from. That must be what they're for. Bats' traguses are, in relation to bats' size, much larger and more finely made than mine.

We are on our way back when a bat passes us, a grayish flitter in our light beams. Dave thinks that it must be a Keen's though now that it is flying it's impossible to tell. It comes back toward us like a yellowish bee, pale in our lights, and then it disappears. It's gone up. It's perched. We go to the place and there it is, hanging: yes, a Keen's. It grins, shouts at us again. Then it drops and flitters away. Now there are two in the tunnel, tacking back and forth. Dave tells me to duck, to press against the wall. He says that they don't always turn on their sonar when they're flying in here, they know it so well; it's like my following of familiar paths and not needing light. He says that they apparently use smell to find their hibernating niches. They don't need sound. If they don't "see" us, then they may run into us. If they run into us, they may get hurt or they may fall and drown. We hunker. They come close and then away, back and forth and back again.

We have disturbed them. It was easy to do.

*   *   *

Dave says that this doesn't matter so much, right now, because it is spring. An off night maybe, but spring; tomorrow may be good hunting. Tomorrow they may all, even the pips, be gone.

At last the bats both pass us and go back, inside, deeper, to their places. Our lights pick out spiderwebs; we are near the end. There are a lot of spiders here. They are orb weavers, fat ones, with egg balls the size of marbles. Our lights attract a crowd of tiny flies with bright green bodies. The flies fly into the webs. The spiders rush in and clamp their jaws, and with a dancelike mechanical shuttling of legs they wrap the flies up. This is a great place for spiders; out of the cold, out of the wind, out of the wet, lots of insects.

Then the air on my face feels strange, it moves, we duck under the phoebe's nest, and we are out.

I shine my light down to turn it off and a fish as dark as coal whisks between my feet and is gone. We splash into the night of trees, leafy slope, lit sky.

It's hard to explain; this sense of opening up, of emergence. It's like coming through a door; this feels like morning, it feels better than morning; the cool, the quiet standing of the trees. In morning I can't look at the sun, but now I can look at all the suns I please.

I think that the bats must feel this when they wake and scatter in the country. They may not see the stars, but our dusk is their dawn; they feed most heavily then, and take a nap at midnight — or several naps — then hunt again just before the light.

They don't have much to worry about out here. It isn't easy for an owl to sneak up on them. An owl's careful silence does not make him any less visible on the screen of their hearing. Owls get a few bats, not many. Bats have a smaller turning radius, can dip and reverse themselves faster in the air. It's like a 747 hunting down a Piper Cub; the little one has the advantage. Some hawks and owls do wait at cave entrances where they manage to pick off the odd one, while the bats' flight path is limited, and while the light is still great enough for hunting birds to see them coming. Raccoons catch bats from time to time, and so do snakes, but often these are bats that have been downed in collisions and, temporarily, have the disadvantage of the ground. Add this all up and it's easy enough to see that predators are not a problem.

Bats live a long time. I didn't know this before. I thought they were like mice or shrews, with lifespans of a year or two. Or songbirds, which live for four or five years, perhaps. Dave tells me now that the bats we have seen can live for ten to twenty years; as long as a horse, a cow. The Myotis bear one young a year, the big browns bear two; their reproduction rate is the same as that of a horse or a cow or — at the most — that of a goat, a deer. What this says in the clear language of such things — the organic language of lifespan and reproduction rate — is this: if they negotiate the glitches of infancy and the blunders of puberty, they will live a long time.

They will live a long time because they have no predators to speak of, their birth caves or wintering caves will not be disturbed, and there will always be insects in the night air.

This is their faith.

Break the tenets of their faith and you can break them. Spreading pesticides with abandon, throwing cherry-bombs in wintering caves to see the manic flitter — these are things that people do. People do these things when they are in the dark; a darkness that is fear, or the consort of fear, which is ignorance. Maybe the darkness is not so grim as it was once. Dave says that there have been rumors among bat aficionados in this part of the country, rumors kept hush-hush because the news is good: bats are on the increase, now, here. Not everywhere, and never fast enough; bats' numbers increase slowly, owing to the limits of their nature; but something has changed.

In parts of the Orient bats are symbols of good luck, long life, happiness. A resident bat is welcome there, like the storks that nest on the chimneypots of Swiss villages, or the bluebirds that accept the offers of suburban nesting boxes. Bats eat quantities of the beetles and moths that feed on forest trees, and the insects that bite us, too; they are as clean and well groomed as the gaudiest songbird and as caring of their young; they look like toys. For all the upside-down-ness and inside-outness of their ways, the time may come when the scrabblings behind the shutter, the little heap of guano behind the rosebush, the dusk drop and flitter, will be pointed to with pride.

Dave hopes so. So, now, do I, though I'm as mystified by bats as I ever was. I understand the lowered tone and the slower cadence of the stories I've heard, and the wide eyes and the head shaking; perhaps I've told my story in the same way, but it can't be helped.

The night is fine as we walk back through the trees, with a morninglike clarity and more silence than the day has, and more brilliance.

I don't watch the roads as we come and go. Dave does the driving both ways. I don't know where that place is, I won't go back, it is buried in the country somewhere, and that is good. The secret is safe with me.

Crypsis

Right now they are out there, too small to be seen, too numerous to imagine.

At midnight I am down in the woods by the river. The limits of sight amaze me still, and knowing those limits I seem to be reduced to insignificance. I cannot really see the stars, I can't see the little things in here, the things I mentioned at the beginning — which are, themselves, both ends and beginnings — so: I see very little. I only see the ragged fringe of forest as it meets the sky.

If I knew only those things that I could easily see, then the world would be a small and unmarvelous place, and I would be big in it: central, leaving marks in the frosty grass, moving large stones, having urgent desires. Aware of the limits of my sight, I can give up all this and be small, free of the burdens of centricity.

The stars are not small. The forms of life invisible to me don't cease to exist. Is this why I like the night so well?

With a telescope one can see the birthmarks of meteors on the face of Mars. With a radio telescope one can, in a sense, hear stars that are too far away in time and space to show a glimmer. With a scanning electron microscope one can look, singly, at the little things hidden in the forest, hidden by the hundreds of thousands: under fallen leaves, pricked into bark. Revealed by nothing as coarse as light, but by bounced trajectories of the smallest of material things, they look like circles of the finest crochetwork, frilly buns of silk lace. They are the eggs of moths.

These aren't just any eggs of any moths. No no. Others are out there, too, but these are these, in particular; they belong to *Catocala,* the underwings. Right now, in April, they *are* the underwings.

One night last August I found one underwing moth on the screen door, a moth driven to dementia by the hundred-watt bulb of the porch light. I didn't know, at first, what it was. I touched it as one would touch the bud of a rose and its color exploded under the feint of my forefinger like a sunset flashed in a darkened room. There was no question, this was it. This was the moth I was looking for.

Since then I have learned more. There are one hundred species of underwings in North America, more or less. They are members of the Noctuid family, the largest family of any of the moths and butterflies. Moths are much more numerous and more diverse than butterflies anyway. Please excuse what must sound like nocturnal arrogance; but this happens to be so. And, among moths, the underwings are an aristocracy of sorts.

I will try to explain why this is.

My explanation will be a work of compression. More is known about underwings than is known about most other moths, so compression is necessary, though people in the know are quick to say that there's lots we don't know, that we don't, really, know much. Here we are; I crank time forward as if it were my hand turning, not the orbit of a planet.

One warm night in early spring — soon now — the lacy buns will shatter, one by one, letting loose green animated threads. The threads will, inchworm fashion, move into the leaves. Before dawn they will press themselves to the margins of leaves; they will look like the margins of leaves. Later when they've grown and molted and are too large to mimic the margins of leaves, they will lay themselves on the leaves' midribs and they will look like the midribs. Later, grown again, they will lay themselves along the leaf stalks. Later still they will emerge from a molt with a new color; gray as bark, with a faint pale frill along their flanks like a fringed skirt and a dark saddle marking their backs. Then, at dawn, they will lie along a twig, or seek a crevice in the bark in which they'll fit like a bead of caulk. The frill breaks up their shadow on the bark; the dark saddle breaks up their length. Together the frill and the saddle destroy whatever vestiges of visual caterpillarness that have been left to them after their color and texture have copied, exactly, the color and texture of the trees. Every day they seek their twig or crevice and lie there, stretched and still. Every night they will crawl back up and out to their business in the leaves. They eat quantities of leaves. In spite of this they do not appear to damage trees too much, they have never been known to cause dangerous infestations, they seem to be in equilibrium with the forests in which they live. The caterpillars grow very large, as long as my fingers, though they are slimmer than pencils. They are as slim as twigs. Their heads look like old leaf scars.

Once the caterpillars grow big enough, they'll pupate, and sometime in midsummer they'll emerge from their private sarcophagi as underwing moths. These are heavy-bodied moths with three-inch wingspans. They are fast, strong fliers. They feed at night on nectar-bearing flowers, sap flows from tree wounds, fermenting fruit, aphid honeydew, anything sweet.

Just before daylight these moths — true to their beginnings — lie doggo on the bark of trees. They are triangular at rest, their forward pair of wings covering their underwings completely. These forewings are variously mottled grays. Some are whiter with darker streaks; these moths rest on birches, orienting their streaks to conform with the birch's streaks. The ones that hide on oaks look like the bark of oaks, and so on.

There are a lot of underwings, as I've mentioned already, and each one specializes in a different kind of tree for feeding in or hiding on. Some are fussy, and won't eat anything but willows. Others are generalists, and prefer maples to rest on though they will take oaks. Some species are so similar as adults that the only way to tell them apart for sure is by looking at their genitalia, or at their eggs. Each one is heir to its own most intimate design.

There are more than two hundred known species of underwings around the world, so there may be a few out there that have yet to be discovered. Discovery is big business among entomologists, since the earliest scientific name is the one that stands, usually forever, with the surname of the namer tacked on for good measure; a flag of immortality. Underwings being as difficult to tell apart as they are, and entomologists as competitive as they are, some moths lug around as many as a dozen scientific names and virtually every species has more than one.

An acquaintance wrote me this:

"Some of the wildest underwing stories of all revolve around the antics employed by the namers to hustle their name(s) into print before their rivals could do so. One (in)famous namer from the last century used to swipe and pin important specimens from everywhere he visited into his cavernous top hat, and trot off to one-up the unsuspecting lepidopterist who just hosted him."

Some underwings have common names as well, and these are easier to use. The common names have followed fashions and some of the older namings are more romantic than descriptive, which gives the lie to the notion that entomologists are either heartless or dry: there is the dejected underwing, the tearful, the darkling, the sweetheart, the sleeping, the penitent, the serene, the forsaken. More recent namings have turned to a mythic style: the Delilah, the Cleopatra, the Titania, the Sappho. Americans in all ages have had a tendency to name their underwings after one another, and the moths are stuck with an immigrantish hodgepodge, like memorial plaques.

Whatever their names are, the moths know to which tree they hold allegiance. This is their real identity: their tree. They find their tree, apparently, by smell. Once they've found it they'll orient themselves — head down or head up in the fashion of their kind — and close down and go still. There they are before light comes, before birds begin to hunt the trees, before jays hop and peer, before creepers spiral the bark. When a bird finds one — and birds do find them — the moths will open up.

Their underwings are shocking: orange and black, pink and black, yellow and black, combinations and permutations of all these, trimmed sometimes with a bright white frill. The bird is startled, as I was, by this sudden explosion of color from what was, only just, discovered to be alive. By the time the bird has finished being startled the moth is gone.

Perhaps it has settled again, closed, on the far side of the tree. Perhaps it's on the ground. Wherever it goes it is unpredictable. Sometimes it has a gash in its underwing, a triangular bit missing that is the size and shape of a bird's beak. The bird has nabbed a mouthful of color, that's all.

The flash of underwing is a last resort. This is roughly equivalent to playing hide-and-seek and hiding behind a half-closed door and jumping out and saying:

"BLAAAAH!" when whoever is "it" is just about to bag you. Then you run, of course, while they're still going:

"EEEEEEEEEEE!"

This is the way it works. Of course, it doesn't always work.

*   *   *

The moths share the night air with bats all summer long. They are big-bodied, tasty moths, desirable bat food, hard to miss at any range.

The underwings hear well. They have tympanal organs in their midsection, the thorax, with which they hear sounds above the range of human hearing. Among other things, they can hear the warning pings of bat-made sonar. Before the bat narrows his beam and ups his volume and zeroes in, the underwings know he's coming and they take evasive action. They go into falling-leaf-like spirals, loops, wild swings, zips, and glides. The bat has a hard time homing in on this erratic jig. The underwing loops up, whirls down, dives for the ground. The bat arrives; the underwing is gone.

Other moths have gone further than this. They can make sounds themselves — a high-pitched rattling like a fingernail drawn across the teeth of a comb, but very high, too high for us to hear. Some people believe that they are jamming the bats' sonar. Perhaps they are creating ghosts of themselves, holes in the air, sound-screens behind which they are invisible to bats. Perhaps they are just bad-tasting, or even toxic, and are warning bats off with sonic coloration, their high whir analogous to a skunk's waving tail, a porcupine's hoisted armory. Whatever it is, bats leave them alone.

Even sturdy as they are, the underwings are not easy to notice anywhere. They are triangular, they look like chips of bark. They aren't elegant or gaudy, like some of the sphinx moths, which also flash colors, which spread their wings to show the pinks and corals and lemons of their underwings and furry chassis. Here there are big black witch moths, with blue-washed velvet wingspans the size of my hand. Luna moths, too. All these are hard to miss. Underwings are easy to miss.

Under the single porch light the underwings are part of the dull crowd bumbling up and down the screen. How many times have I never noticed? There they are. They are drawn by night lights — candles, lanterns, streetlights — as much as other moths.

Big Brown Bat and Underwing Moth
Eptesieus fuscus and *Catocala relicta*

Why this happens is still a mystery. I used to think that moths spent their days flying up toward the sun and that they turned their attention to earth lights only after dusk. This is wrong but it makes experiential sense, and I've heard other people say that they've believed this, too. They half laugh when they admit to it, as if, though they've discarded the childhood conviction, they have yet to replace it with something else.

No entomologist I've met has been able to give me an answer, either. Perhaps, they say, it's a kind of dazzlement. To a moth's eyes, sensitive as they are, even a candle flame must be outrageously bright. Some people think that the region right around the flame looks so dark in comparison to the central light that they fly into that, for safety, and so are drawn around and around as if they were on strings.

Whatever the reason is, the underwing moths can be found on screen doors and windows, in summertime, when they are out and about and when there are lights to draw them. A light trap with a UV bulb and a white sheet or a funnel trap works better. Some moths can be drawn with sweet bait; with fermented concoctions of grape juice, brown sugar, beer, overripe fruit. Every insect trapper has a favorite mix, though they all admit that the longer it sits before it's used the better. A year or two is just fine. A neighbor of mine made a jug of peach bait in 1964 and discovered the moldy-labeled can in his garage in 1986. He decided to give it a try. Underwings loved it. He waters his bait down until it can be applied with a pesticide sprayer — one that has never been used for pesticides, of course. He sprays it on trees: trunks and limbs. After dark he comes back with a flashlight to see what's come in.

The moths come in to bait because they catch a whiff. Most moths, the males especially, have multibranched or furry antennae. These are "nose" organs of no small power. Once the moths are drawn in to bait by the volatile alcohols and esters of fermentation, they can taste the sweetness with the soles of their feet.

They are drawn to other things as well. Most important of all, they are drawn to one another.

*   *   *

When a female underwing is ready to mate, she will find a handy perch and send off a scent. This scent is a highly specific pheromone, an airborne hormone that takes immediate effect on males downwind. Male moths pick it up as far as two miles away, in the dark, across forested country. They follow the scent stream as if it were a road.

It isn't just scent that's specific; it's timing. Some underwings mate only, say, between the hours of ten P.M. and midnight in the second week of August. Others mate only after two A.M. in mid-July. These are examples. Dozens of moth species use the same territory and have the same means of finding mates, and the dividing up — of hour, month, scent — helps to ensure that they do not interfere with others' business.

Once the male — or males — have found the female, they court her. They use pheromones of their own, scents let loose from tufts of wingscales. They may dance, dipping and weaving, flashing color and pattern. They land, touch her. If she is receptive, they will join, abdomen to abdomen.

Moths have wildly complex genitals equipped with patterns of bristles, tufts of scales, teeth, spines. Like puzzle pieces, they can interlock only with the genitals of their opposite number. This is the final insurance. The fit must be perfect.

The joined mating is more or less long. Packets of sperm are passed from male to female, where they will be stored until the time has come for egg laying, for the last heat of summer and the cool of fall; for the death of moths and the long, quiet, lace-bun potential of the egg.

Full circle; a circle as neat as the year, as diverse as seasons. Now, here by the river, I think of circles that weave past in the dark. What else could be so intricate? In spite of the hustlings and puzzlings and romancings of their namers, in what else could we be so unimportant? Underwings may be masters of evasion and deceit, but for most of us, for most birds and bats, too, they do not exist. They are here but they're unseen, unrecognized, incognito. And this, as I understand it, is the meaning of crypsis.

Back Meadow

The grass has come up in the rain, in spite of the frost, a thick dark up-pulse of eager spearpoints thrust into air. The fields have darkened even under this new moon. The grasses are so soaked with dew and with the globular out-spit of their over-zealous drinking of the ground that walking here now is like walking through a pool; I have to wear boots.

There are two meadows here. The front meadow lies toward the road and the back meadow toward the forest. Between the two is a crumbling stone wall laced with briars and vines. One stone has tumbled a little out of the briars. This is my sitting place.

In the back meadow is a pond. The pond is a bawdy-house, a clear eye that has seen too much and is calm, philosophical with seeing. What do you think the gnats are doing there? The ducks? The muskrats? Only the toads and the peeper frogs announce themselves tonight, but more is going on. And has gone on, and will go.

There are other ponds around here, of course, and nearly all of them are claimed by a pair of mallards and a pair of muskrats, too, as this one is, though this one also has a pair of Canada geese. At night the geese go, and the mallards nearly do — they skulk along the bank, bent over, padding silently like old men — but the geese leave here at nightfall. They pass over, cross shaped, low, their wings with a loud *whuh whuh whuh* bending the whole air like a passing freight. I don't know where they go.

I don't know why it is, but I'm drawn here, too. I revolve. I've waited out dawns and dusks and midnights on the stone. I get here and don't go on.

This meadow is owned by the parents of an old friend of mine, and in midsummer when we were children my friend and I used to get old jam jars and come out here to catch the fireflies. They'd begin flashing just after sunset, here, then there, there . . . *run!* In our pajamas and sodden sneakers, we ran. The male fireflies tapped out their traveling call: *dit dit dit . . . daaa,* and waited for a response from a female lying in the grass: *daaa . . . dit daaa . . .* but these were harder to see and catch. Every kind of firefly had its own method; some looped in the air, drawing a luminous J in the darkness, just as we used to do with lit sparklers when we tried to write our names in fizzing vanishing light on the Fourth of July. The fireflies' flashings are codes, mating codes but who cared then, there . . . *dit dit . . .* run!

Now the back meadow has been the scene of more encounters, present ones, no less memorable than that capturing of beetly lights. Just below the meadow is the woods where I held a conversation with the screech owl a week ago. One night in there I found the claim place — claimed by dung left, pile after pile, on a stone wall — of a coyote. Four people have seen coyotes in this meadow, big silent eastern coyotes, larger and more woodsy than their western cousins. One sees them by chance. I wait, hoping I will.

Why do I want to see one, anyway? It isn't easy to answer that. I really don't know. For the purposes of my being here — to sample the creatures that are here on spring nights — I don't need to see them, I know by the dung piles and the tracks I've seen and the reports of neighbors that they're here. I've seen toads before, too, for that matter, and peepers, dozens of times, but something gets better when I see them again. It's like meeting an old friend, but not precisely, since they aren't friends. There is something in the moment of mutual seeing that makes my day — or night — and I don't know why.

From here, the pond is a level oval laid in the canted ground, its surface as clear as mirror. Now centered in the eyelike water is a lone vibration, fine as curved pinstripe — a fallen moth. Two bats tack back and forth against the city glow of the eastern sky, and I see swarms of gnats, too, hung over the water, jigging, and the bats fly through these clouds like fish through bubbled waves. There are slight fillips to the bats' flight tracks as they scoop gnats in their wing membranes or with the webbings around their tails, and duck their heads to take their prey.

Below the pond is one great willow. The long curved and recurved fall of its twigs is a little thickened now by catkins pushing out like fingers pushing, like stippled tongues. The tree still frightens me a little; it used to frighten me more, ten days ago, when I first came here in the dark. Its branches looked too much like arms stopped in clawed midmotion. The fall of its twigs was as fine as hair. It is a big tree. I would turn my light on when I walked in its shadow to check my traps, in which I caught a mother vole and three of her half-grown young. Now the thickening of buds has clouded the branches' grasping angularity and the tree has come halfway to the furring of summer leaves; it looks bowed, veiled, almost demure.

The pond draws me here. Its water is the multicolor of agate, the humic dark of tea, or the glossy reflected pallor of sky, depending on one's perspective. It's set deep in the slant of grassed earth, which feels now as muscled and as full of heave as flesh.

Just tonight the toads came to sing in the pond. Because of cold weather, they are more than a week late. I heard them when I came down the meadow half an hour ago; the coarse ratcheting trill of the American toad, *Bufo americanus,* a song that grew higher, louder, more complex as I closed in. It's a harmonic trill in unharmonized voices, each singer joining and leaving off on a slightly different key so that the whole has the shifting resonance of a chant.

Between bouts of singing there are pauses, and whenever they aren't singing the toads swim around. I can see the dark blots of their heads and the riffles they make. They feel one another's movement as fish do, through clots of sensitive skin cells that sense low-frequency vibration too deep to be called sound; they feel the pressure waves carried through water, the motion of others. They swim at each other in bursts and try to grab each other and the grabbed-at males twitter in protest. I can hear the twittering from here. If and when they grab a female, there will be no protest, but they'll jump on anything that moves, and what they're getting here and now are other males. Catch as

catch can: hence the birdlike chirrups, warbles, the get-off-my-
back mind-your-own-business cheeps.

I go to the north end of the pond where the toads have gathered. Not one of them is singing now, but the water is stirred as
if a netful of fish were being hauled in, seething, full. I want
them to sing again and to hurry them up I try a trill myself,
coarser and not as ringing as theirs, but they're fooled enough to
take my cue and start up; one at first, then two, now five, six.

I watch, crouching down, letting my red-light flashlight
roam. Some singers are not as practiced as others, I see now; one
there has a burry trill that ends, soon, in blurts and bubblings.
The expert singers are careful to prop their front ends clear of the
water, by clinging to a reed or climbing half out on to mud or a
stone, in order to get their resonating throat-bubbles into the air.
The ones who don't do this sound waterlogged. Their trills are
as clumsy as mine.

Well. I have fun with my trilling. When I look up, I see a
semicircle of toad eyes staring into mine. They stare, they swim
closer. They all swim closer. One hops out on the mud at my
feet and looks me in the eye with untoadlike intensity. He seems
to be saying: *But this is serious business.* He seems to be saying:
What do you think you are? I'm not sure, suddenly. I'm supposed
to be looking for peeper frogs tonight. I excuse myself and go
back to my stone.

I'm supposed to be looking for peepers, but I drink coffee
from my thermos instead — I've got all night — and I listen
from a discreet distance to the toads' song and dance.

The toads' arrival is only the latest in a long line of amphibian
debaucheries. This pond is a witness to more bawdiness than can
be easily believed. The evidence is here in balls of jellied spawn:
right now the pond holds the spawn of peepers, the spawn of
woodfrogs, the spawn of spotted salamanders. Soon enough it
will hold the tangled beaded strands of toad spawn, too. The
frogs and toads are obvious in their intentions, announcing their
business loud in the dark (as gray treefrogs will in their turn, and
green frogs and bullfrogs) but the salamanders come and go at

night and in silence. They are big salamanders. The spotteds can top nine inches in length, and you can live here all your life and never see a one.

One neighbor does remember seeing salamanders. She doesn't know what kind they were or the time of year they came, but she says they were big and mostly dark in color. I think it must have been spotteds — if so, it was certainly spring — and they came into her house, a great many of them all at once, and the invasion had the quality of a dream; out of time altogether. The whole afternoon had been rainy and damp. She remembers that. At eight in the evening hundreds of salamanders were on the floor of her study in the basement. As she

Spotted Salamander
Ambystoma maculatum

watched — she was bound for the airport, she was off to a meeting in Zurich and had her suit on and her briefcase in one hand and her other hand on the light switch — more salamanders crept in over a window ledge and fell, damply, to the floor.

It was a visitation, a plague, an omen of monstrous significance; but this woman was of a practical turn of mind, and wielded a snow shovel to good effect, putting the salamanders out — in heaps — on the lawn. She closed the window. She made her plane, and her meeting. According to her husband, the salamanders were gone by morning. Gone-ness is a talent that salamanders have perfected to the state of an art.

Now in April the spotteds are in burrows: crayfish burrows, muskrat burrows, rodent burrows, under leaves, grass, debris, in rotted stumps, in the soil of the woodlands and grasslands: down. Gone. No one knows what they do there. Salamanders seem to spend a lot of time doing not much, though they must eat sometimes, and what they eat is insects; but on the subject of mole salamanders in particular not a lot, as they say, is known. Not a lot is known because it can't be seen. Mole salamanders may be harder to watch than most, though no salamander is easy, since almost every one anywhere in the world is active only in the dark and is not too active even then. The mole salamanders here come up and out on one night a year and it's the night that we are least likely to be out ourselves: in March, in the warm fierce flooding rain of final thaw, the first heavy rain that falls when the air temperature is up above 45 degrees F. That's when they come. That's when they came here.

How do they know when? How do they know where they're going? Some ponds they use, heavily, year after year. Other ponds they don't use. When the night comes, they pour across the country in the dark and the rain, in trickles, rivulets, streams, by the tens of thousands, using regular migration routes. Some of these migration routes cross roads. Some naturalists hereabouts wait for the rain, too, and go out in slickers and with flashlights and flags to stake out the road crossings and steer cars away. The salamanders plod across the tarmac, across the lawns and golf courses and parking lots and through the trees, undaunted by flashlights and fuss or the hiss of oblivious wheels. They come to the pond and belly in. The males court the females, bending, clasping. They are earless and voiceless; they taste one another's presence in the water, they feel one another's presence as the toads do, through their skins. Excited by the females, the males lay clumps of jelly topped with packets of sperm; the females, excited too, take these spermatophores up into themselves. Then the females lay eggs.

When they're done, they go. They don't all leave at once, a few of them may hang around for a week or so, but these are the exceptions.

Sometime in early May their eggs will hatch; in the warmth of early summer their gilled larvae will grow in the pond, will eat or be eaten, and will shed their gills and leave. They will, God and rain willing, be back.

Amphibians in general can be found in unlikely places. I have spent a lot of time in nights past searching those woods below the meadow, and one night I looked under logs and found red-backed salamanders, one to a log, each claiming and defending its underneath domain. They're only as long as my finger, but they lifted their chins and snapped at me anyhow. They lay their eggs in pearly circles under the log and will guard them there until they hatch; they don't need the pond. They eat insects, little ones, insects too small for other things to bother with. Birds will eat them. Coons will, too, and shrews. Red-backeds carry the little insects of the underneaths — dead-leaf burrowers, denizens of rotting wood — up through the chain of being to the mouths of foxes, hawks, owls. Several people have told me that there is a greater weight of red-backed salamanders on every forested New England acre than there is weight of all the small mammals put together: coons, shrews, small rodents, squirrels. Per acre, mind you. I once found clusters of red-backeds hibernating under my woodpile: bagfuls, really. At night they'll come out to hunt. At night in summer they'll climb

trees, and will snap up insects from the undersides of leaves.
You won't see them unless you're looking, or lucky, or both.

Now the toads trill; the toads go quiet. The clouds blow away;
without the clouds to catch and reflect the city glow, the night
darkens. The moon has gone down in the west. It's time for me
to go. I hoist myself now, taking note of my reluctance, again,
to leave.

I cross the high meadow toward the road. My boots drag and
slosh in the sodden grass. At dawn the deer will come here, a
herd of five of them together, bounding over the road and
through the trees with a rush. They'll put their heads into the
grass and whisk their tails. I've seen them do this now half a
dozen times, while I'm sitting incognito on my stone. They're
so blasé I don't care much about them anymore, they're so safe
here that they've gone almost diurnal anyhow, half out of my
ken. By the time they come here the robins and white-throats
will be singing, the pheasants will be sounding off, the geese will
be treading the grass and grazing with their long necks curved
like bows, the toads and the other amphibians will have vanished like a dream. When I walk home through the field then,
the deer will move aside, barely looking up.

A month ago, more or less, the spotted salamanders came
through this field in the dark, parting the grass with their
wriggling motion, their black white-spotted coats glistening like
patent leather; they were careless of all things but the pond. No
wonder that in their delirium of urge they poured so inconti-
nently through the neighbor's window, as if it were a way between stones.

What can I say? Tonight I have my mind bent on movement,
too, across country, boundaryless, in defiance of boundaries,
and it's the houses I'm worried about, as the salamanders
were not.

I want to find peeper ponds. I've noticed that there are peepers
singing in the back meadow pond, even tonight, even punc-

tuated and sometimes overwhelmed by toad trills, but I want to know where else they are singing. I've noticed something odd about peepers. Some ponds that housed peepers aplenty ten days ago have fallen silent now. The back meadow pond was silent then. This is a mystery to me: it seems that each pond or vernal pool attracts peepers in its own time, so that the singing comes and goes here and there over the country. Why?

I can't expect to answer this, I can only satisfy my curiosity as to where. I cross the road and the stream and the next road, and go down into the valley. I know this path now as if it were my own house; the thick sighing pines, the oaks with their rattling vines, the subtle slant of the meadow beyond, the stream that flows with the slightest purling between its stones. Along this drainage some people have built small ponds. The uppermost, the smallest, surrounded by woods, has one peeper sounding off. Lone, rhythmic, it seems louder than a chorus, so loud it makes my ears ring. Across the pond the house has one light shining, high up, as if it were a lantern mostly shuttered. Do the people there hear it, too?

The next pond has no peepers at all. Odd. I see riffles where a muskrat hangs sentinel in midwater. I sneak by. Muskrats, it turns out, are active only at night; except for now, in spring, when they are busy 'round the clock building bank holes and nests.

The next pond is quiet, too. Lawns slope cleanly up the far bank. Two houses, screened from each other by trees, command the high ground. Tonight there are seven deer on their lawns. I count them, their shapes — heads bent, heads reaching up and tearing at twigs. Most of them are just standing around. One is lying down. They look bored, like sheep in a barnyard. Two have their ears up and cocked my way but they don't move.

My own ears prick: I hear it now. Where? Wherever it is it's off my claimed territory, I'm right at the fringe here, but I'm not worried about that. Who will ever know? I cup my hands behind my insufficient ears to catch direction. There: the ventriloquial belling. I move through trees and brush, surprised at the ease of being quiet here, of moving through the woods without a light, of knowing that branches are there before they scrape my face. I'm getting close. There's a house through the trees. There's a pond by this house, too, but the peepers aren't in the pond: no. Where?

I move in slowly, slower; there's a glisten of water over leaves, sticks. It's a vernal pool, a chance depression in the ground filled by a rainfall or two, destined to be empty by summer. The peepers are here. The pond proper isn't more than fifteen feet off, but they're here, not there.

I don't dare turn on the light. I crouch and listen and the belling fills me up as if I were in earphones listening, from some space between eyes and ears, to a music that translates to terminal happiness.

It doesn't seem possible that no one hears, that no one comes, or that no one knows I'm here; this seems so much an event. I'm glad they're not here, I guess. I imagine explaining to this irate and trespassed-upon neighbor why I am squatting in his undergrowth at 11:30 P.M. on a Sunday night. Excuse me, sir, I'm listening to your peeper pool. Would it wash? I imagine explaining that the little frog's name alone is beautiful: *Hyla crucifer,* conductor, composer, musician, orchestra. That they sound off, beginning at Easter, with a copper cross marked on their backs; that they're nearly impossible for us to see or catch; that they consume quantities of insects and are consumed by coons, skunks, herons, ribbon snakes, trout, and bass; that their singing is something else again when you hear it up close like this. Try it, come on; see, it's as solid as bells or violins, flutes; harps plucked a centimeter from your ear. What else? That they announce their mystery, but do not reveal the answers.

I spend the next two hours wandering, happily, anywhere I please. People are asleep, as gone as the toads are in daylight. I've learned to sneak; I like sneaking. I do have an occasional nightmare vision of being surrounded by a wailing gaggle of cop cars with the flash-flash of cobalt light and dark figures with hat shapes and gun shapes talking through megaphones: no, I think. Not likely. I am as invisible here as I am improbable.

Boundaryless, the ground takes back its rise and fall, its gentle drainages and gravel ridges dotted with pines, with oaks. I can taste the woodland underfoot by the sound of the leaves. I can hear water. I can see water by its gleam, hear water by the ringing voices of frogs.

I find three more peeper places. Two are ponds, big ponds. The other one is a vernal pool. In every case I follow the belling in; in the end I have really no idea where I've got to, except that I've arrived. All this time I pretend that my trespassing means nothing to me, but the truth is that the imminence of a confrontation is never far from my mind. It's odd: in this suburban wilderness it's only another member of my own species that I'd rather not meet; the rest is OK by me. Of course, coyotes feel this way, too, and foxes, and possums. In fact, most things do; the notion of boundaries is an ancient and animal notion: it's the others of our own kind we'd rather not meet. Meeting forces decisions of a kind we'd prefer not to make, most of the time: fight or flight are both unpleasant. So: I'm part of the world, too.

Confrontations of the other kind — between species, different ones — are another thing altogether. I think of the toads who stared me down, the masked shrew who explored my palm with the same seriousness I felt in exploring her, the screechy who called back to me and then came, and sat, and looked; in all these cases and in others too, I have had the sense that something was being *said*.

What, then? I'm not exactly sure; I don't know why these confrontations are so good, I only know that they are, and that they're rare. For the most part we have a relationship — with these animals who share our space — a relationship based on mutual sufferance; but when you meet head on it's more than that. What are they really saying?

What the peepers are saying, of course, is fairly clear, and they say it over and over again; my presence doesn't disturb them in the least, if I move in slow and low and am quiet, as you would be quiet coming into any darkened concert hall when the music's going on. Their ringing cry is a highly specific "yoo hoo!" and is applicable only to female peepers, though it's drawn me, too. I squat beside the third and last pond, which is banked by stone and weeping willows, somewhere far to the west of where I began my wanderings. The house above the pond is white and quiet and dark.

I wait, for nothing really; the clouds pass. A lazy wind sighs over the hills. I feel suddenly that dawn must be not far off. It's time to go, I turn east again toward the back meadow. I do not retrace my steps; only later I take time to be amazed that my navigation homeward is so unerring; I don't think about it then, I just go, perhaps that's why. I go straight across the country, skirting garages and flowerbeds, marching up driveways, clambering over walls, crossing roads as though they were bridges, throwing a leg over fences, coming down smack through lawns again and scattering deer like coveys of city pigeons, like rabbits. The deer crash off and go quiet and wait until I've passed. One small dog runs out wuffing along a wall: I stamp at him and make a bearlike "WhuuuUUHH" and he goes off, yipping.

I see this: the houses are dead woods cobbled together, the windows are heated sand, the hinges are boiled earth, the foundations and the ponds are little holes quarried among the roots. The oldest houses are no longer-lived than trees. What lasts? The roll of the land, the moon, the cries in the night.

I come down the home hill on a path. Suddenly then I remember my mother's story. I remember her telling me how, last autumn, in her eightieth year, she came down this same path one afternoon and saw the deer. The leaves were still falling, the earth rustled like tinder, and at first she saw only the does; three does standing beside the path. They were narrow in the loins and their legs were as long and fine as carved sticks, their hooves were like polished ebony, their rumps white and furred like the closure of a sable coat. Then a buck walked out on the path and stood, blocking her way. He looked at my mother from four feet off, his eyes level with hers. His muzzle was swollen and his neck was like copper, polished in curves. The eight tines of his antlers had the luster of old ridged ivory. Their eyes met, she said, with perfect understanding. His look, she said, was a male look. It was full of complicity, power, dreams. It was a stare, she said, of acknowledgment.

I wonder what he saw in her blue eyes. Perhaps the same power, the same understanding, that I saw there when she told me the story.

I go on down the hill and up the next, over patios, through mown fields, woods, through stony hollows where skunk-cabbage flowers curve up, hard as wood, cowled. I walk up stone walls as if they were paths. Is this what it means? To recognize in other animals the raw power that we have not lost?

Now, I think so. I feel I could go on; I could go on forever. A wind has come up and the world is a cool, shifting, silvery place, the trees rattle like castanets. I cross the last road and go down through the soaking grass and I'm in the back meadow again; the toads are still trilling, the peepers still belling here. I sit on my stone. My ears are full of music, music backed by silence — or stillness — anyway, by a deep and lucid dark against which more than music sounds.

It took me a long time to see a peeper sing. Tonight I've only listened, away out there I didn't dare to use a light; but I could go and see it here, now, if I wanted. They are so small, the size of the end joint on my thumb, and narrower. They have curious, alert, upward-looking little faces. They have the rapt stare of a solo violinist who sees nothing but his music. They sit on the bank under a bit of grass, on mud. They blow up their bubbles — they're as big as soapbubbles blown by a child — under their chins. They sit up to let them swell. They pump air back and forth from their bubbles to their bellies, and when it goes from belly to bubble they peep; when it goes from bubble to belly it goes silently, like a breath. When they're done they deflate and sit awhile. In silence. As I'm sitting.

LOUISIANA   June

On every slope a conjuror's charm, at every corner a mystery, in every nocturnal heart a plea for pity, a pang of love, the bitter taste of hunger in the silent mouths, and Exu at large in the perilous hour of the crossroads.

— Jorge Amado, *Shepherds of the Night*

Charon

I checked into a hotel in New Orleans just after lunchtime and then went out to take a walk down Bourbon Street. Now I wonder: what would it be like to dance, nearly nude, in front of almost no one, at four o'clock in the afternoon? With the doors open to the day and the street, and with all of us going by in our shorts?

It's night in that strip joint all the time. Dark, a brownish glossy dark with roseate glows here and there, and a glinting wall, like chainmail or moonlit water, behind the bar. "24 Hrs" says the sign, but all the hours are dark. It's the underworld; a fossorial nowhere, notime, in which one can feel that one will never need to emerge. The man in the doorway beckons to people on the street and says "Hi" to me as I go by for the third time. He is thin to emaciation and is wearing a three-piece pin-striped suit.

After night came I continued my research into the nocturnal nature of man by getting quietly smashed at a bar there on Bourbon Street. By way of excuse I can only say that I don't drink often but that once I've begun I tend to consume whatever is at my elbow until it's gone, and that I have never regretted this.

I didn't intend to drink at all, at first, but the night, like the day, was monstrously hot, full of a moist and cloying effluvium, like a salted steam, an overbreathed vapor. On the street the humidity of surrounding tidal marsh was further cooked by neon light into a soft tumult of lassitude. Tonight (I told myself) was meant to be only a preliminary expedition anyway, a reconnoitering of ground — this street having the reputation of being one of the night-life hot spots for humanity, a resource for which I had come quite a distance — so, I went into a bar to have, I told myself, one beer.

It was a small place. It didn't look like much. But I had heard the music. I had seen the man at the door; it was meant to be.

I found out soon enough that this particular bar had the honor of being the oldest drinking establishment in New Orleans; it had been in continuous operation since it opened its doors for Spanish grandees. Now it featured a blind blues singer with a broad straw hat and a flowered shirt and dark glasses; he was nearly as immobile as an idol, sometimes lifting his chin for emphasis or making a slow gesture with one broad ivory-palmed hand. But his voice! You wanted to crawl into the music and live there.

It was a tiny bar, enclosed like a nut, with nut-brown underwater lighting, tiny overcrowded tables, and a proprietor/bouncer/host with the military bearing and haircut of an ex-Marine. (He was, it turned out, an ex-Marine.) I had christened this man Charon from the moment I laid eyes on him from the street, and Charon he remained, to me, though later he did tell me his name; but it never registered. Ageless and upright, arms crossed on his chest, standing guard in his doorway, his eyes missing nothing either inside or out, possessed of a certain philosophical grimness and efficiency, he was the image of that mythical ferryman who will take one across the river Styx and into the underworld for a farthing's fee.

It turned out that no fee was necessary. Perhaps writers, even in disguise (which I was, in a black dress and without tape recorder or a notebook or any of the other accoutrements), are heir to certain perks; whatever it was, incognito as I was, this Charon made sure that his barkeep kept my beer glass filled and he wouldn't let me pay for anything. Perhaps it was just that I was interested in him. Whenever he wasn't busy he came and talked to me, answering questions that I hadn't even thought to ask. He knew in person most if not all of the neighborhood purveyors of illicit delights and he kept them clear of his door; his was a deeper trade. He knew his customers: the businessmen out for a toot; the honeymoon couples sitting stunned and speechless in their corners; the gaggles of twentyish singles who swept in and out, drinking scotch, their mouths open in a perpetual rictus of laughter; the older women, alone, with hair too carefully molded and lips too carefully red, partial to margaritas; and so on. I had the sense, on my barstool, of being perched on a riverbank. The customers came and went; the blind singer said everything there was to say about the tragic loony delicious mortal world; the beer rose and fell in the glass at my elbow; Charon talked.

At last I asked him:

"Why do people come to Bourbon Street anyway?" because a great many people do come, most of them at night, or for night doings, even if it is the middle of the day. And he said:

"She's like an old whore. All she's got for sale are your own illusions."

I thanked him then, profusely, and left the place unsteady on my feet but secure in my conviction that I had found the answer I had come for, and that this was not an environment which needed any further exploration by me.

I walked through the streets toward the river, the real and actual river, no myth. People strolled by. There were thumps and screams of jazz, barely muted by old walls. In ten minutes I found myself in the nearly deserted quiet of the Plaza de Armas. In the dark this plaza could have been anywhere in Spain or Latin America: grandiose, torrid, quietly crumbling. The Mississippi lay both invisible and inaudible behind its levee, which looked in the dark like any hill. I climbed the levee and then the river curved away black and silent, a barrier. There was no sense there of the river's power, but I suddenly saw — well, I remembered — that the realm of Hades, the underworld of the ancient Greeks, was thought to be a vast marsh threaded by rivers; the rivers of groans, fire, pain, forgetfulness; and Styx itself, the son of Night. The underworld, then. That is what I was after. The bayou-threaded night in all its forms.

Bourbon Street was not a bad place to begin, but the marsh itself beckoned; a huge fringe of continent, sparsely settled and full of life. People go there in the dark, for their own reasons; some go for an honest harvest, that too. Either way, their business then belongs not to the world of illusions, but to the hardest realities of desire.

That some people are nocturnal by profession I always knew, though I was not sure why.

So: in the peddling of illusions I had one answer, and now the marsh itself drew me in the dark and heat. In the days to come I would work with alligator scientists, with a shrimp fisherman, and with nocturnal hunters of other nocturnal men. Charon would propel me in there, after all.

Moon Run

Lake Pontchartrain is just upwind of New Orleans. In the dark you can smell it; rich as bean soup. It is twenty miles in diameter and is connected to the sea by a small umbilical, a channel called the Rigolets; "Regalees," you call it here. The channel is where the shrimp run, like shoaling herring or spawning salmon, in and out from the Gulf — in to feed and grow, out again to breed — and they run in the dark, with the moon.

Even from the air coming in here I could see how richly souplike Pontchartrain was, streaky with cirruslike algal blooms. All the water of the Gulf was thickened with the gray-beige of delta silt and the furry green adjacent land was threaded through with wriggling waterways. There was a flashing of water everywhere: tidal basin, estuary, lagoon, salt marsh and brackish marsh, dunes, beach, canals, levees; and Pontchartrain, round as a bowl. This is the liquid fringe of North America, earth washed down from under the roots of Kansas corn, from under the feet of buffalo, kneaded now by warm and strenuous tides; one of the richest natural environments in the world.

Now the moon is not quite right for the shrimp to be running hard. They are staying, mostly, where they are. That's what I hear; I could have picked a better time, but there are still shrimp to be had, a thin flow, a few, stragglers, the leavings of bi-monthly glut. So just after dusk, after a rainstorm that dumps hard warm rain like plummeting metal, Bill Dekemel and I get into his shrimp boat and push off into the Rigolets. Night falls, a slow shadowing of the high salt-marsh grasses that stand like black ragged walls; now the channel curves and curves in back and in front of us like a country road.

It's like being in the cab of a tractor, locked in with the motor noise. The water around us is as flat and black as a harrowed field. Out there the shrimp that we're after are hidden in the current like nuggets in sand, and the nets that we drag sweep out through sixteen feet and down through twelve feet of darkness.

Bill is deceptively still at the wheel. His eyes flick from radar screen to radar screen, to the lighted compass, to the control panel, to the horizon sometimes, too — to the constellations of twinkles that are New Orleans — and he makes a trifling adjustment to the wheel. He lifts the radio microphone, punches buttons:

"Pooh Bear here. Come-back?" he says, and is answered by a crackle. He grins.

Only two other boats are out tonight. It's the half moon, so the tides are muted and what tide there is is nearly slack, and we're not expecting much. When the moon is new or full and the tides are strong — the moon's pull compounded by the pull of the sun — there are nights when more than a hundred boats work this channel at once. Then the radar screen is solid with boat streaks blurred like comet trails, it's a trick to keep from colliding, the ship-to-ship radio waves are crammed with hailings, a three-man crew is kept busy sorting shrimp on the work table and packing shrimp in crushed ice. Three or four basketfuls makes a good night. We'll be lucky to get one. On a great night you might get twenty. Once a season, maybe, you'll get a hundred and fifty.

Last year, Bill says, more than 39,000 commercial shrimping licenses were sold in Louisiana. Shrimping rivals even oil as the number-one industry of the state. All this human business moves with the moon; the moon is to nocturnal sea creatures what the sun is to land: prime mover of season, clock hand, rhythmic source.

Bill started shrimping in a small way twenty years ago. Nowadays he's the president of the East Bank Commercial Fishermen's Association, a director of the Commercial Shrimpers of America, a member of the Governor's Task Force on Shrimp, an adviser for the National Marine Fisheries Service. His boat is big, powerful, hi-tech.

He's a foursquare man with square hands like a farmer's, he has blue eyes that seem too blue to be true and a thick boyish thatch of honey-colored hair. Somewhere (one doesn't wonder why) he was dubbed Pooh Bear, and to keep things simple *Pooh Bear* is the name of his boat. His son's boat has *Pooh Bear's Son* painted on its stern.

This is, he says, real shrimp country, the heart of the whole deal. From Florida to Texas the whole Gulf coast is blobbed with estuary and brackish lake and all of this is nursery ground for shrimp, he says, though some places are better nursery grounds than others. Shrimp (and what they feed on) are fussy about salinity and temperature, among other things. Hurricanes and storm tides can pull the plug on a good season; shrimp are fussy. They're fussy about what the bottom is like, because they need to hide in there during the day. They like sandy organic mud that they can burrow into and they don't like bottoms that are coarse or shingly. Pontchartrain, he says, is just right.

"It produces a larger shrimp inshore than any other inshore body of water on the Gulf coast," he says.

The pink, white, and brown shrimp come in here from the sea when they're less than half an inch long, tiddlers, nearly translucent; the color and size of thumbnail. They move in by the millions with each rising tide. They bury themselves on the bottoms to wait out tidal ebbs and the brightness of the days, then they come up and move forward again in each night tide until they're in. They come quite a distance for what they are; some come as many as a hundred miles just to get to the Rigolets.

The adults spawn at sea between May and July and the larval shrimp come in here late in June to feed and grow. The season for catching them on their way out again runs from May to November; the early catches are shrimp that came in the summer before and neglected, for one reason or another, to leave: they're monsters when they do leave, a handful each. Shrimping is so lucrative a trade here that sometimes whole families move onto their boats for the summer, dogs and cats and all, with playpens in the cabins.

Once the tiddlers are up in their nursery grounds, they feed in the estuary gumbo and molt often and grow fast. They eat polychaete worms, tiny crustaceans, algae and fragments of seagrass, mollusks and tunicates, diatoms and fish larvae, anything in reach. Most of the live things they feed on are nocturnal, too. The shrimp are grazers, hunters, omnivores. When things are right, they can double their weight, Bill says, in twenty-four hours. They may stay in here for three months, or nine months, and when they're done with growing they wait for the moon. They wait for dark and a steep tide to swim with the flow down the Rigolets for the sea.

This is where the action is; this is where you catch the jumbos running for the open Gulf.

They swim by gliding forward on a rippling of abdominal swimmerets, their tails stretched out behind and their feet tucked in. Their stalked eyes jut up toward what light there is; the water-wrinkled prickles of stars, looms of land, horizon glows, the moon. These round stalked eyes are densely black with sensitive rods, and red reflectors curve in back of them, so that what light there is can be doubly seen. They do not run blind.

They have their predators, naturally, but they stand a better chance of getting by the speckled trout and catfish and red drum in the dark. When they are chased or confronted, their great defense is the backward skip, and they put most of themselves into it, into the bound and coiled muscle that is the edible part and is tipped with spreadable flanges like a fan. If you come up to a gliding-swimming shrimp from anywhere, the fan spreads and the muscle contracts and with a leap she is gone into blackness; pursue and she leaps again, zigging and zagging in dizzying flings.

"Ever tried to pick up a live swimming shrimp?" Bill asks.

"No."

"It's like trying to hang on to popping corn the size of a kid's hand and as slick as an eel. Heck! It just about can't be done."

Shrimp have wiry antennae that spring from their faces and surround them in a heart shape, trailing along their flanks and tails. These antennae are long; the white shrimps' are twice as long as their bodies are, the pinks' and browns' are a little shorter, and in water they can sense movement so well that it's hard to believe; it's as if we trailed stiff whiskers ten feet long, a kind of force-field. They can sense current; the direction of its flow and their own within it. They can feel the bow wave of a fish, a boat. The antennae can smell what's there; they know the salt of sea, the brackish stews of Pontchartrain, the nature of threatening approach.

Bill has laced his craft with sensory apparatus, too. Hunters have to take not only the place and the rhythm but the technol-·ogy of the hunted, or some of it; an approximation. There is the radar, which can let him sneak so close to land he could touch it

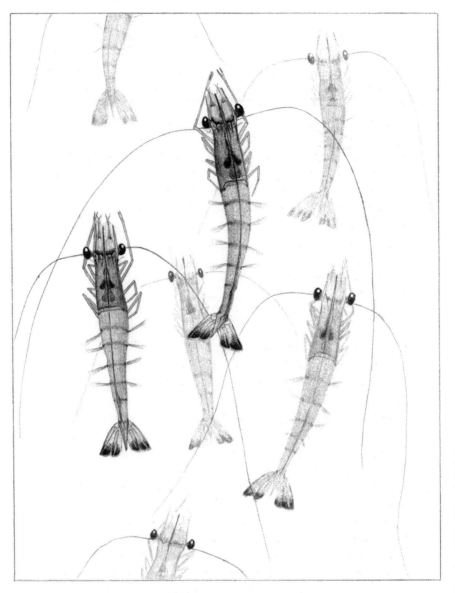

White Shrimp
Penaeus setiferus

with a stick; and the side soan sonar, which shows him exactly where his net is in relation to the shelving bottom. He has fuel gauges, RPM gauges, a compass, an autopilot, color video sounder, depth sounder, a radio for communications. Outside, the nets are hung on two rectangular sixteen-by-twelve-foot metal frames that are unfolded like butterfly wings from overhead and then lowered to each side. The nets are attenuated cones that trail far beside and behind us, underwater, like the antennae of the shrimp.

"Right then!" he says. "Ready?" and I nod.

He pushes levers and turns the wheel and the motor noises change, and we're out of our chairs and outside. Outside it's hot, the air thick as steam. Winches turn and the net booms rise into the sky and at last the nets' "tails" come up, bulging, and in the deck lights the bulges twinkle red like the brake lights of cars seen far away: shrimp eyes.

The net tails are pulled inboard and Bill yanks the drawstring that holds one shut and a mass slithers wriggling, popping, into a bucket. We spread the bucket contents on the white table under the lights, a long slither of silver, rose, black. Flexing rose-white curves of shrimp are in there, and other things are in there, too.

Olive-green crabs scuttle and clatter and scooch behind the baskets, eyes goggling. Tiny white puffer-fish pulse into textured balloons. There are big-eyed silver fishes that are as thin as paper and as long as eels, and other fish, small and large; they flip and bounce across the table, fins erect. We throw these back and they arc overboard, hit water, disappear.

Bill's hands are fast; here, there, into the bucket, overboard. Around us the night is black, a velvet curtain; here the spotlit white table slides and dances under our hands; there is the smell and feel of clean sea, a slime on our hands like oceanic soap.

This is less a harvest than a sampling. Fully half of what we've caught is shrimp and most of these are jumbo whites longer than my hand. There are ten to fifteen of them to a pound. They're as big as they come, the width of my wrist. They flex in my hands when I pick them up, snapping like metal. Under the flanges of their armor their bodies are wonderfully firm; tensed muscle. We toss the jumbos in one basket and their long red antennae tangle in our fingers and break like threads. There are smaller shrimp, too, which still qualify as large. They go in another basket.

"Look here," Bill says, "see the color?"

He holds a shrimp in one palm and points; though the body of the animal is pale, the swimmerets and the legs and the borders of the shell are pinkish red.

"They're not too deep-colored, see," he says; "when they're really running now they're deep-colored, oh, deep red all up there. In the day they're just pale."

"Why?"

"I don't know."

Out here on deck the thick air tastes of sea and rich earth; up there one airless planetoid, moon, draws atmosphere and oceans; the moon syncopating with the sun to make the rhythm of the sea. When the shrimp are running, Bill, of necessity, goes nocturnal, too. His day begins and ends with the night tide, at two A.M. or dawn, and still nothing of this is something you can ever, exactly, set your watch by, he says. Some harvests are just about miraculous; others, for no known reason, thin. Now the drawstrings are tied again and the nets flung out and the winches slowly lower the nets, and we move forward, the motor groaning with the weight of water filtered through; the marsh grasses, dark walls, sway. The stars are not immutable either; nothing is precise, ever, anywhere.

This is the fringe of ocean, or land, you can't say which. They are woven together here, two fringes twined. The boat is spangled with running lights; there, behind us, another boat runs spangled, too. In darkness, light becomes language: as we go on fishing I think of the deep-sea shrimp who have more than a hundred light organs each, on eyestalks and carapace margins and abdomens and legs; they light up nose to tail like the flashing ring of lights around a movie marquee. Some of them, alarmed, spit luminous clouds. Though they live in the deeps between six hundred and thirty-six hundred feet, they move upward toward the surface to feed, every night. How do they know, I wonder, that it is night?

William Beebe, who spent a great deal of his life pioneering the ocean deeps, discovering organisms that no one had ever seen — animals whose light organs are used for camouflage, for seduction, for hunting — wrote this in *Half Mile Down:*

> Whenever I sink below the last rays of light, similes pour in on me. . . . The eternal one, the one most worthy and which will not pass from mind, the only place comparable to these marvelous nether regions, must surely be space itself, out beyond atmosphere, between the stars, where sunlight has no grip on the dust and rubbish of planetary air, where the blackness of space, the shining planets, comets, suns, and stars must really be closely akin to the world of life as it appears to the eyes of an awed human being, in the open ocean, one half mile down.

Mystery abounds; even in the Gulf shrimp, known to almost everyone, hunted by so many and eaten by millions, mystery abounds.

The sea, even at its edges, is less underworld than otherworld, companion to our own, hidden by surface, dark by nature. Here at the sea's fringe the moon gives a cycle to run by; these things are familiar to fishermen more than to anyone else.

I think of the California sardine fishery that did business only in the dark of the moon. Schools of guzzling sardines stirred the night water up so much that pools of phosphorescence formed where they fed, and fishermen wrapped their nets around pools of glimmer and hauled — not light, but sardines by the ton.

I think of the palolo worms of the South Pacific reefs. The palolo worms mate only in the third-quarter moon in November (a few mate in the third-quarter moon in October, jumping the gun), and it's done by proxy. Worm bodies full of eggs and sperm snap off from the head (which stays, chaste, in the coral) and the snapped-off proxies wriggle to the surface shedding eggs and sperm whenever they brush together. They can be scooped up by the boatload, roasted in breadfruit leaves, and they taste like shad roe. Lunar caviar.

The Hunt at Sabine

Craig Guillory is a Louisiana game warden. He drives his truck fast, he reels off the wildlife refuge rules like a catechism, and pulls his hat down firmly over his nose.

The rules are simple: no firearms, no liquor, no bows and arrows, no illegal drugs, no commercial gear. Tonight we're going to the Sabine, a wildlife refuge in the southwest corner of the state; two hundred and eighty square miles in the heart of the coastal marsh. We're almost there.

Craig is slim, thirtyish, gangly, relaxed. He makes me laugh — he has a gift for understatement and a naturally comic way of putting things — but I wonder what I'm doing here. He wonders, too. He tells me he doesn't like the idea of having a reporter along, and a reporter from the *East* . . . and a *lady* reporter from the East; and then he grins and pulls his hat down and spits out of the window.

It's late in the afternoon and the sky is dark with squall clouds. We drive fast on the straight roads over the flat country.

He calls me "Ma'am." I call him "Mr. Guillory." The soft, wet, late June heat barrels in the window, and we are polite.

The reason I'm here is that Johnnie Tarver has let me come, let me into the marsh, really; though I didn't know, at the time, what he'd let me in for. Johnnie is the chief of the Fur and Refuge Division of the State of Louisiana Department of Wildlife and Fisheries, and I told him that I wanted to live in the marsh and I told him why; I was writing a book about nocturnal wildlife and people. Johnnie greased all the wheels. Two hours ago I moved into the big guest house at Rockefeller Refuge, which is a vast chunk of native salt and freshwater marsh; and, among other things, Craig's base of operations. Now I'm here in this truck; and though neither Craig nor I are sure why, Johnnie was sure enough.

"Those game wardens down there are night *predators,*" Johnnie told me. "That James Nunez and his boys down at Rockefeller are the best. That's where you have to go. We'll get you mixed up with them and then you'll see something," he said.

My timing, though, is bad; James Nunez had heart surgery a month ago, and nowadays he's confined to his desk. As soon as I settled in, I went to see him there.

The wardens' offices lie just across a grassy lawn from the guest house; outside in the sun the heat settled on me like cloth, an enormity of heat. The marsh stretched away, a blinding furred green, threaded with glints of water; here and there egrets stood, on poles, by the water's edge, as white as coils of ice. The air was rich with a smell like fresh-baked chocolate cake, laced with a faint oceanic tang, a wild sharpness. Again I had the distinct idea, almost a vision, as I gasped across the patch of lawn toward the wardens' offices, that this blue of sky and green of marsh were illusory things that would rise, with the dusk, to reveal . . . what? I didn't know.

The game wardens would take me there, I knew that. Johnnie had promised me that.

James Nunez is slim, dark-haired, ruggedly handsome, and doesn't look a day over forty though he must be nearer sixty; he has the same shy gentleness about him that Craig Guillory has, a quality almost of softness, which is — I would find this out — the most deceptive thing in the world.

Craig and the other game wardens who work for James work all the time, seven nights and days a week, he told me, when I asked. "Hours?" he said, as if the notion hadn't occurred to him before. This is no profession; it's life. War, of a kind. James made it clear that here, in marsh country, poaching used to be a way of life and game wardens used to be a joke; they are not a joke now.

"Children 'round here used to be weaned with the caution always to keep an extra cartridge in their gun, for the warden," he said. "Some of them still do."

James has handled more than 25,000 cases in his career to date — his own estimate, and he's a modest man. He told me right away that he's done most of them at night. All the important cases were done at night. He said that nighttime is when you catch the real outlaws, the criminal offenders, the professionals. That's when you catch them on the back roads, the highways, the canals, the bayous, in the marsh.

James's predatory heritage is clear enough; his first rule of work is that you have to know your territory. Then you have to know what crime is likely to be committed: gill netting in the canals or bayous; Q-beaming, "jacking," "bulleyeing," for deer; duck poaching after hours or over the limit; baiting pools for waterfowl; butterfly netting for shrimp in refuge waters; smuggling: what, when, who. Know where it's likely to happen — just into the bayou mouths, at the weirs, on the open marsh. Know the poachers' patterns; when they'll come, where they'll go, the routes they're likely to leave by. Set your stakeout, collect the evidence: see the deer carcass placed in the pickup. Watch the garfish stacked like cordwood. Then: "Pop 'em."

James has never refused to catch anyone, he told me, and he treats everyone the same when he does catch them. He's caught most people in the county at one time or another. Nowadays, he said, the old "I'll-poach-you'll-smuggle-nobody's-gonna-tattle" network has begun to reverse itself. Much of the wardens' information now comes from people they've convicted.

"The old-time outlaws now want people caught," James said.

Hunters and fishermen were the first conservationists in the country and are the strongest of them now; James knows this for a fact. His best informants are the best hunters. They're the first to notice when something is wrong. They know the territory, too.

*　　*　　*

Later, when we went outside, James showed me his car lights. There's a panel of switches under the dash. He can switch off his brake and interior and running and head lights and turn on a tiny lamp on his bumper, like the ones the army uses, a veiled slit, and with that he can drive seventy miles per hour in the dark and no one can see him coming.

"If a man says he's a warden, ask to see the switches in his car," James said. "If he doesn't have switches he's not a warden."

James is known in person or by reputation all over the southern tier of the state: "James down at Rockefeller," "that James;" not just in Cameron and Vermilion but in St. Mary, Terrebonne, Lafourche, clear from Texas to New Orleans. The reputation he's earned isn't just for himself but for all the wardens he's tutored and marked with his creed, everyone he's worked with: "James and his boys."

"I wish I could take you out there myself," James said, lifting his chin toward the marsh, the vastness of sighing, glinting green. I wished so, too.

"You'll be going out with Craig tonight," James said. "He'll show you around."

I asked him about Craig. I'd already met Mr. Guillory out by the truck, with his hat on backward, his gunbelt on sideways, his arms full of rain gear, an old pack, rifles.

"Craig's a country boy," James said. "He's comin' along real good."

Craig tells me that he was born and raised in Cameron Parish and has never lived anywhere else. Like most people here he's French by descent, and his is the first generation of the family to speak English as a first language.

"My grandpappy didn't like the English much," he says. "He always said that if they wanted to come down here to Cameron Parish they'd better learn French."

Perhaps it's meant as a warning off, though it's said in his same soft way, like a joke.

We drive through Oak Grove and Creole and Sweet Lake. We pass fields of beans and rice, rough cattle pasture, then wild grassy ground and open water; the marsh has a smell of sun-baked wicker chair. It seethes with green and light.

"Over there is Locum Ridge, over there is Hackberry Beach," Craig says. "In between there's nothing but marsh."

There are no towns here, he tells me. Locum Ridge and Creole and Sweet Lake are streaks of habitable dry ground. Six years ago they put a traffic light in Creole and it's the only one in the parish. The parish government is all the government there is. A sheriff's office and a few local deputies patrol, but there are no municipal police, he says: the wardens have full police powers. They enforce the laws of the wildlife refuges — there are three big ones in Cameron: Lacassine, Rockefeller, Sabine — but the wardens have done all kinds of law enforcement — DWI work, petty larcenies, murder investigations, drug interdiction.

This is a country mostly empty, flat marsh threaded with veins of bayou. This is part of the drug runners' gold coast that stretches from Florida to Texas; country that is open, empty, soon to be dark.

"Just got houses on the road," Craig says. "You seen a town? That's it — wa'n't no town."

Before Huey Long put roads in here, back in the forties, he says, there were no roads. There aren't many now. What few there are run the ridges that parallel the coast, and that's where the houses are, lined along under the moss-bearded live oaks with here and there a tighter cluster, a corner store: Pecan Island, Grand Chenier.

The ridges were Pleistocene shell beaches washed equally by the sea and by Atchafalaya effluvium; their soil is oyster shell wadded with continental runoff. They're muddy shingly dunes left high and dry in the continent's postglacial rebound. Craig points them out across the flat ground of the marsh; darker lines of live oak, palmetto, hackberry. In the old days, he says, in a drought, you rode a horse across from one to the next and took the paddlewheeler from Cameron; that was the way out. Fur trapping, farming, 'gator hunting, that's what you did.

Craig wears a short-sleeved green-brown shirt with blue warden's patches on the shoulders, and his long arms are pale. It's high summer. His pallor is my first clue to how much his profession is night work, to just how much he is out there in the dark.

"They use the cover of darkness," he says, "and we do, too."

We've skirted the Sabine and are heading north.

"Where are we going?" I ask.

"We're going up to Grand Lake to put in," he says. "The important thing is, always to make them think you're going someplace else."

Them. They. It comes to me suddenly that we're after someone; people, identifiable particular people. There are men out there that we're after, that we're going to *get*.

Until that moment I'd somehow managed to think that I was just going along for the ride, for the inside tour to see the wildlife; alligators maybe, deer, nutria, night herons. Now I know. I am going on a hunt.

Now I know why he doesn't want me here. I don't know if I want to be here.

We pull into the Speedy Tote, a corner grocery gas station. Another warden, Malcomb Agbert, is waiting for us there. The boat is on a trailer behind his truck, the hull sleek and gray, an outsized 150-horsepower Mercury outboard heavy on the stern. Malcomb is dark-haired, broad, handsome, his warden's uniform blue. The wind has risen, the sky is dark and billowy with cloud; we go in and buy Cokes and stand around in the fluorescent-lit aisles. Lee Benoit, the local deputy, is in there too, and a young woman who is introduced to me as the local cub reporter.

"We got one judge in Cameron and one cub reporter," Craig says. Everyone laughs.

Craig buys supplies for the trip: a six-pack of Gatorade, two packets of chips, six pull-tab cans of Van Camp's Beannee Weenees, two tins of Spam, a large bottle of Rolaids.

We watch the storm. Toward the sea the edge of the cloud is like torn charcoal-gray paper. The hackberry trees churn and the grasses lie over in the wind. At last we get into Malcomb's truck and pull away in a spurt of gravel and head north. We take one road and then another, each smaller than the last, until we pull up in the gravel by the lake shore.

There are a few small painted houses here, a wooden pier. Water laps at the foot-high shoreline of marsh earth that is the color and texture of chocolate, braided and woven through by grass roots. Out of the muck rise tall tasseled stems of oyster grass, salt meadow grass, black rush; taller than my head, the grasses seethe and sigh like corn. Two gulls climb over the water and go seaward with a purposeful slide that looks to me like homeward arrowing; they're going toward their night place to bathe, joust with their kin, squabble, sleep. An egret slants by beneath them, laboring against the wind. The far horizon is grass, peninsular juts of marsh are more grass; it's a planar country, fluid, the water as muddy as gumbo and the marsh as watery as potter's slip; a country as devoid of marks as desert or sea. Flow is all: wind, water, grass. The wind is strong and the lake surface is whitecapped, too rough to launch into, but it's early still. It's still light.

Craig changes his clothes piece by piece behind the truck, swapping his warden's uniform ("they're always after us about the dang uniform") for camouflage gear: pants, shirt, jacket, camouflage hat. A pair of camouflage bandannas is tucked in his pocket. A gun goes in the boat, and my pack, and his pack, and the Beannee Weenees and the Gatorade, a poncho, a coil of rope.

Malcomb has changed from his uniform into camouflage, too. He emerges pulling his hat down against the wind, tucking in his shirt. Craig hands me an old green combat jacket ("you'll need these long arms against the 'skeeters, Ma'am"), then

Malcomb gets into the truck and backs the trailer and lowers the boat down the ramp. Once in the water she looks lively, eager. The wind has died down now and the whitecaps have gone, and the lake surface is streaked with late light. We get in, the Mercury roars, and we're off.

The motor noise is deafening and the boat bangs and jackhammers over the chop as if we were sliding down a boulder slope in a plastic sled. When I stand up and look over the windscreen, the wind mashes my nose and yanks my bandanna and the banging is so severe I have to be sure to keep my tongue out from between my teeth and stand on my toes, like a deer, taking the jolt in my ankles. It's easier to sit. They stand, hats on backward and chins out, shirts flattened to their chests. I don't know then but discover later that they're on watch; standing, they can see just over the grasses to whatever lies nestled, to what shows over marsh and levee. The shoreline is featureless and dark, a ragged fringe. The rose of dusk comes and goes. At last we curve toward land. The darkness opens to a streak of bayou, and we go in.

Suddenly around a bend there are three roseate spoonbills perched by the water. We slow and slide toward them like a skier skidding downslope; the birds rise and fly off, low, shockingly white and rose. We grin at each other — the birds' uncanny beauty has jolted us, for a moment, together — then the Mercury roars again and we go on.

Now and then the water widens into a pond and we turn toward one bank or the other and always the shore opens into a lane of water. There are no landmarks that I can see, nothing but marsh grass and this maze of watery bayous, the Indian word for river; but these are hardly rivers; they run everywhere like tapering veins in the soft body of tousled grass. Then there is another pool and this time we curve in toward the grass and I catch my breath because there is nothing there ahead of us *but* grass, the higher silhouette of a grown-over levee and no opening at all; and there in mid-pool the Mercury roars in complaint, we slow to a stop, the water around us thickens like chocolate milk.

"Dang. Bar!" Craig says.

Mud. We tilt and back away, churning muck; we back off until we're clear, and then rev up — the men holding their hats — and we run for it; the Mercury screams and spews mud and water and then we're through and sliding into the grass with a whish. By some miracle there is water under us, the grass is over us and all around in a rush; it's like diving into a cornfield; this cut through the levee is no wider than the boat. We drift forward, pushing the thick growth out of our faces. We come to the end suddenly — there's a wide lane of water ahead — and there we stop, rocking, hidden, having come in the back way. Craig takes the painter and jumps to the bank.

"And the other thing is," he says, softly, talking to me over the bows, his hands busy with the rope, continuing the conversation interrupted hours ago, the one that began with *make them think you're going someplace else.*

"You got to know your enemy. Know the place. And the other thing is," he continues, "be downwind. Upwind voices carry a long way, especially across water."

"You see them big Texas boats back there?" Malcomb asks.

Craig nods. "We'll get them after," he says, and there's a flick of his eyes, to me, to Malcomb, a sudden knotting of tension, a hardening, but they say no more.

Craig gets back in the boat.

"You got 'skeeter dope? Right, then. Grease yourself up good," he says.

"Them 'skeeters are going to be *bad*," Malcomb says.

I can hear the thin whines, see the hovering shapes against the lighter sky, feel the tickles on my bare skin. The whines thicken to a chorus. Coached by Craig, I empty half a bottle of DEET over myself, soaking my hands and face, spraying my clothes. We spray each other's backs. It's easy enough to tell what you haven't sprayed enough; patches of pantleg and shoulder blacken with settling swarms. I spray my hat. I soak a bandanna, too, and tie it on Western-bandit fashion over nose and mouth, copying them. Craig soaks a second camouflage bandanna and puts it on his head so that it hangs down over his neck and ears and

back, then he puts his hat on over that. When he stands up in the
bows, his night glasses to his eyes, peering down the bayou, he
has the silhouette of a French Foreign Legionnaire.

Something has shifted in him with the dusk. He has straight-
ened, become purposive, serious — though not all serious. But
he has changed. He is in charge.

What we do is wait and listen. We can hear motor noises,
faintly, in the distance. There's a ship channel at the end of the
bayou we've come to. There are lights moving in the distance,
too; how far off they are is hard to tell in this featureless flat. The
men tilt their heads, listening.

"They're in here for sure, but how far in?" Craig whispers.

We wait for them to come far enough. We drink coffee from a
thermos and open a can of Beannee Weenees and pass it around,
using our knives. Craig gives me a 'skeeter switch, a long horse-
tail of finely shredded palm fronds bound together and tied to a
wooden handle.

"Made it myself. From palmetter. In the old days wa'n't no
'skeeter dope and that's what you had," he says. "That's all my
grandpappy had."

The switch works; it keeps the ticklings off my face, the swish
of the fronds is softly scratching, cooling, pleasant; it matches
the sighings of the grasses around us, it gives me something
to do.

This is the hard part, the sitting. James said so, now I know
so. Moonlight silkens the water, cone-heads and crickets cheep,
rasp, shrill; the chorusing calls of these countless insect voices
seem, now, in my nocturnal fancy, to be the twinkly chirrups of
stars. Mosquitoes whine; they've divvied the night hours as
moths have, into blocks of time, each species making its rounds
in turn (some do bite in daylight: a minority stance), but dusk is
the communal high point of their busyness. Different species
bite at different times, anyway, though it isn't something
you'd notice particularly in the dark. I remember reading that
mosquito-borne diseases — yellow fever, malaria, encephalitis,
and so on and so on, it doesn't bear thinking of — are carried

around in the insects that like to do their biting in the early morning hours when they can find us surely napping, the DEET worn off, the 'skeeter switches motionless by our sides.

So: we sit for hours. I begin to fall asleep but mosquito pattings and whinings rouse me whenever I begin to nod, and anyway there's no comfortable place to wedge myself, so when Craig stands up and says, quietly:

"Dang this. I'm goin' down the levee after 'em,"

"I'm coming too," I say.

And there's no comment, so I go.

The grass is thick underfoot and it's like walking on a bed. We walk the water's edge where the sedges and seacane are short enough to see where we put our feet, but after half a mile the thick growth of the levee comes down to the water and we've no choice but to go up. The cordgrass rises overhead and there are thickets of mangrove and salt cedar that are eight and ten feet tall. Craig pushes through and I struggle to keep him in sight. There are invisible branches underfoot, holes, stubs, deadfalls, mashes of vegetation. Suddenly he stops.

"Them 'gators 'll hear us comin' and get out from under," he says.

"I hope so," I say, not having thought of that before.

"I hope so, too," he says, and he doesn't sound too sure.

Alligators are nocturnal; at least now, in the heat of summer they are; and I realize that I've only ever seen them in daylight, torpid in a zoo, inanimate as algal mud that has been modeled into sodden, if classically horrific, shape. They can gape and rush, of course. I don't really want to see one now. I don't have time to think about it, anyway. Keeping up with Craig is enough.

The good thing is that the mosquitoes leave us alone when we're moving, and there's enough of a breeze up on the levee that they're down to bearable even when we do stop. We stop often. Craig holds up his hand and then we stay balanced to make no noise at all while he listens. The night glasses do him no good in this high brush, and the image intensifier — the heavy black cylinder that looks like a zoom lens and magnifies existing light into a greeny video image — is no good here, either. We are limited to ears.

We hear motor noises. They are hard to locate, ventriloquial as night noises are. They bounce, amplified by water, muted by grass. I cup my hand to an ear and I can find them better now, still ahead, and separate one motor from another; big boat rumbles, small boat pockety-pocks. This is why, I think, foxes have the ears they do.

Craig turns to me and whispers: "They *know* we're not in here." He grins, and I can see the white gleam of his teeth.

We don't stop long and we keep looking to see if the water's edge is clear again, if we can get down out of the thick growth, but it isn't and we can't, and the bushes thicken. My bandanna slips, my hat's snatched off time after time by flailing salt cedar. We stop again. He listens, shakes his head, turns.

"I'd hate to think how many cottonmouths I've trod on up here," he whispers.

"Cottonmouths?"

"Yeah. Bet you trod on a couple, too."

On we go. If I were alone my old snake horrors would make me helpless; cottonmouths don't like the daytime heat any more than I do. This is their hunting time. I put them all, mentally, down in the wet ground hunting for frogs, out of my mind.

I don't know how far we're going and I won't ask. There's no sound but the popping of crushed seacane and the creaking of branches. A bird rises suddenly from the marsh below us, a dark long-winged shape rising with a *whop whop* of wings and gliding off, tilting, settling. A night heron. We stop as it rises, go on when it settles.

It's nearly midnight. When we stop again, Craig listens, nods, turns to me.

"The only thing we haven't come across in here is a hornet's nest," he says. "If I run back over you and keep on goin' you'll know what I hit."

I begin to giggle then, the one-liners and the long march having got to me at last — alligators, cottonmouths, *hornets* — all real enough, but I can't stop the giggling and it comes as no surprise somehow that he's giggling, too; we're wiping our eyes, hanging on to branches to keep ourselves upright. . . .

When we've subsided enough to move we go on, but not far. We go no more than fifty paces and then he stops.

"You hear that?" he says.

"Yes."

The motor noise is coming, flanking us, down, to the left, just ahead now: there it is. They're in.

"Let's go," he whispers.

We plunge down toward the water but there's no way of telling where the water is. We are going fast and I trip and fall and then he does, and we're up again and pushing our blind way through and down. He whispers:

"If I go in head foremost you pull me out, hear?" Then he begins to giggle again and I do too, helplessly; we choke ourselves to silence just in time.

We see the gleam of water just in time and catch ourselves, hanging on to branches, and we crouch there, the pockety-pock noises loud now.

"They *know* we're not in here," he whispers again, seriously this time, and it's true; I know it's true, and I nod, pressing my hat back on. No one should be in here.

Here they come, here is what we've come for: two dark boat shapes, one of them and then another, black silhouettes with not a light shining.

"Shrimpers. They got their nets down. They're in the refuge. No running lights. Yeah," Craig whispers. He takes a deep breath.

He has his radio in his hand, he punches buttons.

"Malcomb." A crackle. "They're in," he says, his voice as

low as the sibilance of grass, the slipping of the little waves at our feet. "Come and get 'em." More crackle. "We're way the heck in the puckerbrush. You can get us out after."

The shrimpers pass. The wait is long, we're hunkered, still; the silence and dark press in on us, the grasses sigh in the wind, and the whole marsh seems to hold its breath; then there he is, the little boat slicing up like a blade, the light shining, the bullhorn.

"Game warden. You're under arrest. Pull up your nets."

Malcomb's voice sounds neutral, soft, almost bored. He's past both boats, rounding them like a sheepdog. Craig talks him into our patch of bush, the hull rattles in against the branches.

"Jeez!" Malcomb says, ducking. "You picked *puckerbrush* all right. Heck!"

Then we're in the boat and off. We're dancing over dark water again, like herons risen into air; into stars again, the float of the wind; I take a deep breath. Two, five. The bayou is shivering planes, the levee is a dark bulk fringed against the night sky.

The shrimpers have stopped. Their motors putter at idle. I see now that there are two men on each boat and there's a boy on one of them, too, nine or ten years old maybe, out with his dad. The boats are not much more than flattish platforms with rough square cabins, a heavy array of butterfly booms and winches; they look cobbled together, rusty, slapdash. There is a light on over one of the sorting tables and a man is working there in overalls and shirt-sleeves, head bowed and hands sorting the shrimp into white buckets. Handfuls of waste fish flip glittering into the water, we hear the ticking of crabs sidling on the floor-boards, and no one looks at us. There is a sudden smell of fish, a whiff. The boats are dead in the water, the winches groaning, the nets rising against the sky.

"Just trying to make a livin'," one voice says out of the dark.

The nets are up now and the heavy ends are out of the water, dripping. Pairs of shrimp eyes glow a deep bright red there in the beam of our light. The illicit catch hangs, globular, black planets with red neon-lit cities pricked all over them. Thousands of shrimp, thousands of other things — fish, crabs — scooped from the broth of night water.

"Empty your nets," Malcomb says through the bullhorn.

"Into the water?" the voice asks, softly, knowing the answer already.

"Into the water," Malcomb says.

There is a jerk as the drawstring is pulled, a ragged splash, and they're gone.

We all go out toward the ship channel and tie up there to a jut of grassy ground. The shrimp boats' running lights are on now; so are ours, we're like a lit village in the dark, we tie our boats up and come out onto the grass. The shrimpers squat or stand in their overalls, swatting mosquitoes. Craig is there in the middle, looking more like a French Foreign Legionnaire than ever, while he reads them their Miranda rights.

"Just trying to make a livin'," one man says again, when Craig has finished.

The voice is the same; dulled, bemused. One shrimper is still busy at his sorting table; he's allowed to keep whatever he caught before we caught him. Craig and Malcomb climb onto the boats and begin to take down the nets. The shrimpers help. Craig hands his thermos of coffee around. I offer my orange bottle of 'skeeter dope, and the men take it and nod, and the boy takes it, too, and nods. He has, exactly, the bowed shoulders of the men.

It's over then; the citations are written out and the nets are rolled in neat green bundles and dropped into our bows. The nets will be impounded for four months and both sets are nearly new; each set is worth, Craig tells me later, more than eight hundred dollars. Each boat owner will be fined fifteen hundred dollars. It's heavy damage, I think, though later Craig tells me that in a night or two of shrimping refuge waters — nursery grounds where the channels are narrow and the shrimp run thick, protected from other fishermen — thousands of dollars can be made.

So I think of the boy, and I think of the boats cobbled out of green rough-sawn timber, and the shrimp running under us. We watch the stripped boats putter down the channel to home.

"They were friendly," I say.

"We're all kin here," Craig says. "Kin stealing from kin, kin arresting kin." He grins, then sobers, presses his hat down, backward, so the wind won't flip it off. Then he turns to Malcomb.

"Let's go get them Texas boats," he says.

Malcomb nods, and again their eyes flick. The knot of tension tightens. Malcomb guns the motor and we turn, sliding, sliding again, and skim back in between the narrow walls of black waving grasses into the marsh.

The Texas boats are not kin. No one is sure what they are, though we have, all of us, a common suspicion. The men stand straight, silent, as the boat zips and slides again through the bayou maze. There's plenty of drug running going on all up the Gulf coast; and these are big boats, seagoing boats — they've told me that: "Heck, they got no *reason* to be in off the Gulf in the first place, unless . . . heck!" It's so easy to pull deep behind a levee to do shady business; for the wardens, of course, it's tempting to think that that's why they're here.

"What are them Texas boats doing up here in Louisiana?" Craig says now. "You tell me."

"They're not lost, I promise you," Malcomb says.

"They know where they are. They know where they're not supposed to be," Craig says. "Dang. And we got no warrant. All we can do is ask them to leave."

There they are, suddenly, large and pale against the grass. Two of them, sitting nose to nose against the levee in the dark. Big sea-going commercial trawlers, their masts bulging with the pods and dishes of navigational gear. Craig and Malcomb are quiet. We tie up to the stern of one and the men turn their hats around, tuck their shirts, settle their gun-belts on their hips. Before he leaves Craig pulls a double-barreled shotgun out of the locker, breaks the barrel, and hands it to me.

"Ever used one of these?" he says.

"No," I say.

"You do this," he says, snapping the barrels into place, "and this here's the safety. On, off. See?"

I nod my head.

"Don't for gosh sakes pull both them triggers at once or you'll blow yourself overboard." He hands it to me. "'Bye, Sis," he says.

They go up on board together, shoulder to shoulder like French gendarmes, like city cops.

They're gone.

I sit down low so that no one can see me. So I'm Sis now, I'm not Ma'am anymore. I allow myself a glimmer of pleasure at this, then I feel the gun heavy across my legs. I duck down deeper under the thwarts.

The sky is the brightest place, a colorless brightness; the marsh is as dark as coal, like the fur of bears. The shrimp skip by underneath me, I think, now, and the strange long fish that are as flat as sheets of hammered silver, with huge dark eyes. The crabs rise, scuttle, flail down there in the long sloshing pull of the moon; the underworld, I think. Well, I'm in it now. In it now. Is this Lethe, the river of forgetfulness? Cocytus, the river of groans? One of countless capillary fingerings: bayou penchant, bayou misère? The gun is heavy on my lap, sensual and utterly foreign, cold as water.

I look at the stars and there are so many of them that I have trouble picking out any ones I know. I'm distracting myself, or trying to: Draco the dragon, Perseus with the head of Medusa in his fist . . . I hear knocks in back of me, on the boat, overhead and then a door opening, and a voice, a man being noxiously overpolite: "Thank you, sirs, for letting me know . . . yes, of *course* I under*stand*. . . . I am so *grateful* . . ." I hear the wardens' footsteps, military and precise, cross to the other boat and I hear them knock on that other cabin door. I hear them go in. They're gone. The man's voice behind me shifts then to low angry mutterings, curses; there are slams, clatters.

Then there are soft footsteps coming up from below, sneaker feet tiptoeing across the deck, a pair of voices nearly overhead, two young men.

"They gone?" one says.

"Let's cut the motherfucker loose," the other says. "See what they fuckin' do then. Cocksuckin' bastards."

I don't want to be cut loose. I can feel the tide ripping through underneath — from where to where only the gods and the wardens know, but I'd get there fast enough; and I know even less about outboard engines than I know about double-barreled shotguns, which isn't much; so I stand up in the light and click the barrels home with what I hope looks like a practiced swing.

"What are you doing?" I ask, hoping my voice is soft, bored, a warden's voice.

"Nothin'," they answer.

One is standing, the other is sitting on the post to which we're tied, the painter in one hand, a knife in the other. They're very young. I don't think of it then but I remember afterward what Craig has told me about kids, about scared kids being the most dangerous of all.

I don't think of it then; but I do later — how I must have looked to them popping up from nowhere with my squashed hat, straggles of hair, red bandanna awry, in a ragged oversized combat jacket and with mud to my knees, like something dragged up from the marsh; chased out by hornets, maybe.

They don't move. The knifeblade gleams.

"Get away from there, now," I say then, and they do. I sit down again but so I can be seen, so the light gleams on the barrels.

When Craig and Malcomb come back, we wait while the trawlers rev their engines — they're low-thrumming, powerful, like truck's engines. We wait, bobbing in the current until the boats move off. We follow them out to the ship channel and toward the sea. We putter softly through the dark chop, in their wakes. The air is cool; a black deep drink of sea wind.

"How'd you do?" Craig asks.

I tell him about the boys.

"Cuttin' the boat loose, that's tamperin' with police property, that's a federal offense," he says, softly. "They woulda gone to jail for that. You did good, Sis."

He doesn't sound surprised; I'm not surprised either, any-more. I'd follow Craig now, I realize, to the moon. He'd let me come, too, maybe.

The Melting Land

The next day just at dusk I get into a boat with Craig and Junior Vaughn, his partner for the night. This time, Craig says, we will stay here at Rockefeller and run the canals looking for gill net-ters. Gill nets can be strung from shore to shore, tied to small stakes driven into the muck; the nets will catch and entangle anything that swims in the dark: nutria, mullet, garfish by the boatload. The gill netters are sneaky quick, Craig says, but our boat is faster than theirs. We bang and slide down the long zigs and zags of canal, kicking up a rooster tail like snow.

The sky colors and darkens to a deeper translucent blue, the chop is slivered with silver and rose like battalions of fish scales. The first stars prick out, mosquitoes thicken the air, tiny hail against our faces in the wind. The chorus of insect voices rises in the grass like a million small screams; the marsh smells like warm skin, a smell of salt. The tasseled grasses billow as we pass. The men are intent as we travel toward the sea, watching the levee edges to see if a boat's wash has wetted grass and earth above the waterline, watching for stakes driven in and twine tied; listening for motor noise.

Halfway to the sea Junior throttles us down to idle and we slide in against a wooden jetty, an adjunct to an oil platform. Oil platforms dot the refuge, oil companies once dug and now maintain the system of canals; oil revenues support Rockefeller's enviable system of marsh preservation, protection, and research. Craig beckons and we get out and squat on the jetty. We soaked ourselves with 'skeeter dope almost before the boat slowed to a stop; the insects settle in immediately, like the heat. Craig puts his finger to his lips and points; a dark ragged shape floats just there, low in the water. He purses his lips and lets loose a series of squeaky barks; the 'gator turns toward us and comes slowly in. Two more appear, closing in, too.

It's difficult for me to tell how large they are. Their eyes are above water and so is the bulge of their nostrils, sometimes, and if you know what you're seeing the rest is not too hard to estimate by the length of the head. A few saw-toothed scutes appear, too, at the neck. Black bitsy silhouette is all there is; the rest of the animal lies, iceberg fashion, under the surface of the chop, giving nothing away. They turn toward us and drift in, as a log would turn and drift in an eddy.

There is method to this drift; 'gators have sensory pores around their lips that can feel and taste the most delicate pressures of motion in water, as fish can, and fish is part of what the 'gators are after; so, motionless float is ambush. So is low profile, since the swimming mammals — nutria, coons — see very well in the dark, and they won't launch off if they see anything too untoward out here. Never forget a 'gator's tail, a bundle of muscle that, lashed, shoots the animal through water like an outboard; or the legs that tuck, streamlining the body like a flattish sword; or the forward end, with a gape like a barrel and teeth like tusks. These are meat eaters and no mistake. Later I read that the biggest waterways — here, the canals — are claimed, defended, patroled by the biggest bulls.

Craig yelps again, sucking air, and the silhouettes shift, align, approach; Craig grins. What he's making is the distress call of a baby 'gator. It works like a charm. As coyotes and wolves are drawn by the agony scream of hares or deer, 'gators are interested in distress of any kind, hoping to bag the distressor or the distressee while they're otherwise occupied. Sometimes the big ones may nab or at least scare off whatever is bothering the baby 'gator and the baby will escape. Sometimes the baby will be nabbed, too.

Well; one comes close enough below us so that we can shine

our flashlights and see its bright owlish eyes, very wide, and its odd frontal frown and simultaneous grin, like a demon clown's, and the size of its back, armored like a dinosaur's.

"A nine-footer. Maybe ten," Craig whispers.

The tiling of its armor is the deep black-green of night water or deeply shadowed grass. It shuts its eyes then and ducks and with a lazy wave of tail swims off: nothing interesting here. We get back in the boat and go on.

At last we reach the sea. The wind, unsifted by grass, comes in off the Gulf burdened with tropical moisture. It's fully dark now, the last of the sun's glow gone. We've come to the end; we stop and tie the boat and get out, stems popping under our feet. The marsh muck has the texture and smell of new-plowed ground. Living in this, I think, is like living in black and breathing basketwork. We squat there and smoke and talk, in the sea wind and out of the 'skeeters, and they lecture me about the warden's trade as if I were an apprentice to it now:

"You got to put yourself in *their* shoes, Sis."

"When you out there in the marsh at night, alone now, you think of Taitais . . ."

"Taitais are Coonass boogymen. You scared, see! When you hear a racket at night, might be a hog or a dog or me or you or Junior here, well, you turn and shoot!"

"Especially kids."

"Especially kids, like I told you."

"So you watch yourself."

"You hear a gunshot out there, in the dark, you don't know who he is out there."

"If he hears you out there he squats down and can see your silhouette."

"You get down too, see."

"If he lies down you can't see him. And he stays there, see. But you know where he is, and you walk up to him hoping it's not a scared kid or somebody spaced out on dope."

"We've made coke cases in duck blinds."

"Most of the time you go to a complaint area and sit, wait, listen for shots. With field glasses you can see the fire from the barrel. With a night scope it looks like a yard of fire, like a light saber, and a cigarette lights a man from the waist up."

"People are easy to see."

"A duck hunter is sitting, listening for wingbeats. You have to crawl through the marsh."

"Creep 'em, like stalking a deer or something."

"Then you hit them with the light beam, blind them, say: 'Game warden. Stop. Put down your gun. Stand up with your hands up.'"

We open sodas and drink, with the stars and the lap of the water.

"That 'gator back there, now," Craig says. "He thinks he owns the place. Rightly he does. We aim to keep it that way."

"And the garfish, he owns a little of it, too."

"That's right."

"And the coon, he don't own but what he can steal, like us."

They laugh.

"Anyone who eats so much crayfish can't be all bad."

"Coon's like us wardens, reachin' around the place in the dark."

They laugh again.

"Cajun's what you folks call us, we here call ourselves Coonasses."

"This is our life. Right here."

"Heck. My purpose is to make a difference," Craig says. "Maybe my grandkids will have something to fish, something to hunt. Maybe your grandkids will."

"If there's any marsh left," Junior says. "Let me show you, here."

We walk toward the sea and underfoot I feel the grasses grow thin, they're wadded and crushed over, then gone; all that's left is a dark mudflat with nubbins of rootstocks and a bush with branches torn away, one tuft of leaves straining in the wind, a desperate silhouette like something left after a fire, a bombing; then the sea is there, planar and black and whispering, flecked with starshine.

"It's washing away, see. Right here," Junior says; he's almost shouting in the wind. "Right here. Next storm this will all be gone." He waves his arm, taking in the mudflat with the vanished grass, the marsh beyond.

The muck underfoot is as slick as mayonnaise, there are no lights on the water, the waves make a noise of wilderness.

Fifty square miles of Louisiana is disappearing every year, Craig says. It started back in the forties and fifties, not so you'd really notice. Now whole islands and coastlines are washing away. You'd notice it now, anyone would.

The side effects of the oil business — the canals dug to bring in and maintain the rigs — bring the high tides and storm tides in deep; salt water comes up into freshwater marsh, killing the grass and trees. Then there's the waterway, Junior says; it runs the width of the state from the Sabine to New Orleans. They say that's changed the drainage patterns that brought silt and fresh water down to the coast here. They've roped the Mississippi in with levees, that too. When the river had its way, it cut new channels for itself, spewing silt and water over the landscape like a garden hose turned on full and let loose to fling itself. Now, one third of its silt, Craig says, is captured by headwater control structures. The Army Corps of Engineers have stopped the floods; they've stopped the rebuilding of marsh. And levees guide what's left of the river out to sea. Eight to ten thousand cubic feet of silt-laden fresh water flows into the Gulf every second of the day and night. Two million tons of sediment a day is dumped off the continental shelf. What would make land sinks into the abyss.

The land is melting here like chocolate left in the sun. Perhaps it always did melt but was rebuilt by the river. Maybe marsh is one more thing that we believe in our ignorance to be as permanent as stone, and are finding to our surprise to be alive, and mortal.

They tell me this, and this is no enemy you can stalk after dark, no gunfire you can watch for; fifty square miles a year may be stolen by the Gulf of Mexico; but there is no ticket to be written out, there is no justice to be done, not easily, not just by risking your skin.

We are quiet for a while on the slippery flat, the once-marsh already half chewed by sea, then we leave. We run the canals again, for hours, and the men climb levees and sweep the country with binoculars and night scope, but we see no one and hear no boats. There are no lights except for the twinkling of the oil platforms, and once we pass the rumbling rusty bulk of a dredge at work, keeping the canals open. The men wave at us from their lit cab, and we wave, smiling, back at them.

Notes from Bird Bayou

2 A.M.

I came out to Marsh Island this afternoon in a fast little warden's boat. When we left Point Cypremort, it looked at first as if we were going straight on out to sea. Then there was a long darkness on the horizon, suddenly, like a pencil line, and this thickened gradually into a frayed wall of grass. With a wedging in of glistening water the grasses opened, in the now familiar way, into bayou.

"Bird bayou!" shouted the warden at the wheel, as we sliced in out of the sea, into a sibilant and surprising calm. There, also suddenly, was a white house on stilts, into which I was ushered; into air-conditioned cool and every comfort; in which I have been ever since.

All of this island is wildlife refuge; one of the most remote pieces of salt marsh anywhere. The annual alligator hunt starts here tomorrow (today, really, isn't it? It's past midnight: yes), and there's a gathering of the clan here; wardens, naturalists, scientists, fur and refuge administrators; Johnnie Tarver himself. I'm a guest, an interested person, to whom all opinions can be addressed and by whom all data can be readily absorbed. I have been kept busy. These men know more about alligators in particular and marsh in general than any group of people anywhere, and we've been sitting at the table, talking, for hours.

So: I'm here to find out about alligators, among other things, and this is the right place; but suddenly now I need a breath of air. I get up and find my boots and the stairs down and out, and then the screen door slams behind me and I'm into the heat and the stars.

Here I'm out of the hum of air-conditioned cool and the smell of spicy crab-boil and cigarette smoke and into the night air, which is as thick as wool, moist and heavy as a towel. The sky is immense; strewn with precise pricks and fainter cloudings of stellar design. I feel the bump of the 'skeeters and hear their whines closing in on the giveaway come-on that wafts from me like scented flame. Here the air holds the familiar baked-wicker hot-chocolate scent of marsh, and sudden cooler pockets of sea air. The water at my feet has a reptilian glitter: opaque, inscrutable, only water.

Beyond this narrow water, marsh is the world. House, boat-house, lawn, jetty, are an acre and a half of civilized lozenge wrapped in twin arms of bird bayou. Beyond this is island — eighteen miles long and eight miles wide — a chunk of pure salt marsh veined with water and wrapped by sea.

If Louisiana is a boot shape, then this is a fleck of muck kicked off the instep arch. There are five miles of wind and chop and the pulling tides of Cote Blanche and Vermilion Bays between here and the nearest point of land. To our south is the Gulf of Mexico.

Marsh Island is a plain name, which suits the place. There is only this one house here; the headquarters for the refuge wardens. The house is ample and square and white on its government-issue stilts, its flagpole stands empty like a warning finger facing the reach to the sea.

I feel the old panic rising: if marsh is what's here then how am I to *get* to marsh — by what oblique means? How can I get to the life here that goes on at night? Not only the night of bourbon and fists slammed on the table and data logged in and diagrams drawn; but the swimming of the alligators, the feeding of the nutria and the deer, the hunting of the snakes, the fish which come in to feed or breed, the poise of night herons by pools. If marsh has seduced the men in there then how can I touch the thing myself? Or perhaps it's seduction that I'm after: some confluence of the senses that can own the soul?

I want this: night like Pandora's box. The diamond clasp undone, the lid wrenched back, the demons loose.

I have, I realize now, a kind of hunger, a thirst, for darkness, for the imminence of the strange.

Strangeness abounds. Here and now I want to see 'gators, and this isn't unreasonable: by anyone's estimate, there are thousands of them here.

Here the 'gator seems to be, forgive me, a kind of god. Top of the food chain, big, once threatened (perhaps; opinions vary) by extinction and now come back into its own, it's the standard-bearer of the marsh. Now in the heat of summer it's fully nocturnal; like snakes and other reptiles it shifts its allegiance according to temperature, rather than light.

The hunt begins in the morning, the annual harvest of Marsh Island 'gators. The hunters have camps all down the bayou; in the morning they'll go out in their boats to see what the night has brought. The baited lines have been hung since evening. Where, now, in all this, are animals closing in?

I leave the jetty and turn back, pursued by a flurry of questions I haven't asked, and then by prickling surety that this may be a world too rich and big to know at all.

The night air is charged with the brimstone scent of the mythic, with the slip and slop of water that is like sibilant speech. I look down the bayou once again, but there is no sign now of the hunters' camps, just one red glimmer of a fire dying down where the water bends away, where bird bayou disappears and becomes one with the line of grass.

8:20 P.M.

Dave Taylor is an alligator biologist. He's drawn a sketch to show me how he catches baby 'gators at night. It's here in my notebook, like this: there's a pole hanging off the front of a penciled airboat and a stick figure is suspended from the pole by a string. The stick-figure is labeled GIRL.

Alligator bait. The 'gator hunters, I know this, use bits of chicken or beef lung. They hang this bait from a line that is wedged into the end of a cane, the cane holds the line over water, the other end of the line is tied to a stake driven into the muck. The 'gator will come at night and take the bait, in which is hidden a hook; once hooked, the animal can be pulled in next morning and dispatched; but what we are after tonight is the live and unharmed creature. Any bait, I suppose, will do.

I can tell that Dave doesn't want me to go. It isn't just the drawing that has made this clear. I put on my hat and boots and he puts on his hat and boots and rolls his eyes. A young warden is coming with us and he is excited; he grins, his eyes gleam, he bounces down the stairs. Dave tromps down, sighing loudly and shaking his head.

"For Christ's sake," he says.

Outside it's dark, starry, hot, calm. We do ourselves up with 'skeeter lotion.

The airboat is perched, teetering, on the edge of the lawn, its bows over water. It looks like a flattened metal rowboat with a huge fan mounted amidships; there are two seats high up, in front of the fan; otherwise — except for a small space in the bows — the boat is filled with and surmounted by metal struts. Dave and I get into the high seats and the young warden holds on below. The sudden roar of the engine communicates itself through our seats as massive vibration, as whole-body hum. We slide forward into water and move away.

We're out; the grasses half below us now, a tossing heave in our wind; the stars shiver overhead as if blown, shaken. We skim down bird bayou and Dave shines a heavy-duty flashlight,

American Alligator
Alligator mississippiensis

a night-blaster, over the water; a bazooka of light. The light is, somehow, brown. The water is suddenly transparent: we see clear to the bottom, to the brown silt. The air is smoky with damp and the grasses seethe by, and there . . . *there* and *there* . . . twin round sparks shine back at us, bright as answering flares, magnesium-bright: 'gator eyes. 'Gators are everywhere, suddenly, all over, near the grass, in the water. Almost under us paired torches sink down and extinguish, and then we're over the animal and can look down into water as if it were smoked glass and we see him swimming: a six-foot 'gator with his front legs tucked under his belly, his tail flailing him down and his webbed back feet splayed to steer him away.

We grin at one another like pilot, navigator, gunner grinning on a bombing run in an old World War movie: they're here.

The Leviathan of Job; Sebek the water god of the Egyptians, who roused the Nile to flood every year; Typhon of the Greeks, the monster with a hundred dragons' heads, father of lethal winds; here they are.

Their eyes are knobs above the water, their nostrils another knob, the scutes as rough as bark or shadow. When we pass the darkness closes, they're gone. Then there are more: here, look, a big one; the eyes glow so bright and are big, far apart . . . gone. He's gone.

Right there I have a sudden vision of a movie seen too long ago and too late at night to have absorbed much more than a lurid generality: a dinosaur lurches from a swamp, screaming and raking its claws in the air. The dinosaur is hot on the trail of Errol Flynn (or someone); anyway, its colors are snaky-oozy, its tread bomblike, its eyes filled with liquid Hollywood fire; and our hero (half naked, naturally) has every hair on his head slicked in place.

The dinosaurs have this powerful tug on us, they're the quintessential monsters: mindless, terrible, huge. The American alligator is almost the only one of the old saurian cousinage left around; this is what I'm thinking. And this is where he lives. Take a map of North America and blank out everywhere there's frost, everywhere a glacier's been, and everywhere that isn't wet, and what you have left is alligator country. Wetland fringe: soft ground washed onto continental shelf and well furred over by green.

In Texas and Montana people have found fossils of a monstrous 'gator kin. Fifty feet long with a six-foot skull — its forelegs as thick and high as you or me — it hid in marsh water, lurking like a sunken galleon. Its name was *Deinosuchus* and its prey were dinosaurs.

Now there are only little ones left. The American alligator isn't the only one. There are twenty-three species of crocodilians alive now: the American croc in southern Florida, the Nile croc in the Nile, caimans in South America, the gharial of the Ganges, the little Chinese alligator in the grassy valley of Yangtze Kiang. That's all of dinosaurs that's left us now unless you count birds. Birds — the fossils point to this all the way — are descended from some dinosaurian branch line that transistorized components, overheated its core, and took to the air.

We're birdlike now; flowing over, high over the water, seeing it all down there as if we were owls, like a movie run fast-forward. A little 'gator — little for canals — four feet maybe, maybe a little more, climbs out into the grass, swaying like a lizard, quick, and goes; the tail-tip gone, the grass closing. They're fast; I've heard this, now I see.

'Gators like to eat birds. A 'gator hunter told me this. Dinosaurs still eat dinosaurs, it seems. He said he'd use blackbirds or grackles for bait now, if he were allowed, if it weren't against the refuge rules, and he winked. He'd hunted 'gators all his life, they'd put bread on the table back when he was a kid, though not as much as they do now. Back then they were worth a couple of dollars a foot. Now, this year, they'll bring more than thirty dollars a foot. He'd done some poaching, too, in days gone by, though you didn't consider it poaching, really, back then.

They'd go out at night, in summer. They'd take a light and a gun.

"That was outlawin', they called it," he said, and winked

again. "Their eyes glow in the dark, you know, just like coals," he said, "just like coals. And we'd cut the motor and sneak up, and shoot . . . and grab 'em quick before they sink."

"Sink?" I said.

"A dead 'gator sinks like a stone," he said. "Just like a stone."

That was work. This, baiting them to a hook, is easier.

"But birds now," he said, "is what a 'gator can't resist. It's like a kid with candy."

A blackbird that perches for the night over water is daring the gods. A 'gator will rev up scoot and rise and snap . . . or whock it into the water with his tail. In brackish marsh they pull down gallinules and mottled ducks. Dave and the other biologists have told me that they choose their meals according to their size, the time of year, personal taste, and the place they find themselves. They'll take mud turtles, cottonmouths, watersnakes, catfish, garfish, mullet, sunfish, shad, sheepshead minnows, crawfish and blue crab, shrimp. The little 'gators back in the marsh will root for insects and tadpoles, rubbing the mud with the sides of their heads; perhaps they can taste what's buried, perhaps they can sense life there as sharks can, by the presence of the electromagnetic fields that are given off by all live things. The big 'gators who lay claim to the deep waters of bayou and canal eat mostly meat: rice rat, mink, rabbit, armadillo, raccoon, possum, muskrat. Birds when they can get them. Most of all the big ones here eat nutria, because nutria are plentiful and they swim at night.

The 'gators, so I hear, are equipped for eating anywhere anytime. Their eyes are more light-sensitive than ours, thick with rods, and with their slit pupils they can close their eyes like doors when the sun is bright; they can open them wide in the dark. Wide, with reflecting tapeta that shine back at us like lamps. Their nostrils are slitted, too, to close under water. They have a valve in their throats that can be shut so that they can feed even when they're submerged. . . .

Now there's something else, the light just catches it, up there in the grass, a humped shape.

"Nutria!" Dave shouts.

Then there's another, by the water, with a chinless rodent face and beady eyes; it dives into the grass.

They're big, not as big as beaver, twice the length of muskrat; more than two feet long when they're grown, not counting the ratty tail. They're not native here, so I've been told, they're native to South America. They were brought in to populate fur ranches and some were let loose and others escaped; they've pioneered all this. Like beaver or muskrat they live in burrowed bank holes, build grass nests, and feed at night on vegetation: sprouts and grass roots. They breed like . . . like nutria; as many as three litters a year of up to eleven kits each (mostly it's six or eight), and the kits swim and eat grass when they're a day old, they're ready to start breeding themselves at five months or so; even given that what they eat is grass they sometimes eat more of that than the marsh can stand. Nutria "eat-outs" reduce rich growths of black rush and three-corner grass to shredded nothings, and then the soil washes off and what is left is water. People point this out: "That pond there was from the sixty-eight eat-out, right there. Heck, you could walk from here to Oyster Lake on the backs of the dang nutria!" No one begrudges the 'gators their share.

Far off I see the grasses quiver — coon, is it? 'Gator? No; a lolloping nutria, another, three young in tow, crossing a piece of marsh that is as flat and dark as a bearskin rug. The light swings away; they're gone.

A pale mist rises from the water and swirls as we skim past; up ahead, part of the starred sky is blotted by thundercloud, lightning flashes, the thunder bumps. Ahead three pairs of 'gator eyes shine out of darkness, like lightning focused.

Then we swerve, sidle, the motor hum sharpens to scream and our bows meet the grass of the levee and we tilt back, and back — I don't believe this — we go up, tilting as if over a wave, and the airboat twists and bellies over the levee and down, slowly, sidling, gradual as a slow walk, and we're over out of the water and into the wide plain of shorter grass: this is where

I've never been. Here the marshland is wide, horizon to horizon. We creep forward. Suddenly there are deer in front of us; a coppery doe and a coppery fawn that rouse, slowly, from beds in the grass. They trot away with their ears back, unconcerned, their dark noses to the air; they stop just feet away and stare after us and then we're past, moving over grass stems bending like a wheatfield.

We're into the backcountry. Our lights pass and cross, beams like searchlights; outside of them the night is thicker, blacker than it is when you don't have lights. Pools of water lie here and there in the fieldlike flats. We come on them suddenly; they smell bitter, thick, primeval, their water dark, their shorelines scummed with algal growths. A night heron rises from one, flies past, pale and long.

Then suddenly there's a 'gator nest in the midst of the grass like something left from a haying time; a mound of dry brown stems lashed and dragged and heaped last spring by a mated female. Dave has told me that this nest will take her a full week to build; once it's built, her eggs are laid in a crater in its top and are covered over and guarded until hatching. Generally, the mother 'gator rests with her chin on the heap, as if it were a pillow. The heap's heat of decay and insulating size will keep the eggs as warm, or cool, as they need to be, and the tolerances are close: if the egg incubates at 89 degrees Fahrenheit what will emerge will be male; if it incubates at 87 degrees, a female 'gator will hatch. If you're a 'gator, then what you are depends on your heap, your place in the heap, the angle of the summer sun, thunderstorm and tide, the shade of the mother 'gator's head. We pause there and shine the lights but see no mother 'gator near; still, it's not a place to get out; if she's still close she won't want us around. We glide on, twisting over grass and water.

Then there's another pool, larger, shallow, rusty brown. A scoot of color and a wink of light: there, there's a baby at the edge of the water, one of last year's hatchlings, just a foot long with coffee-and-cream colors bright as a lizard's, nose small and cute, eyes huge. Herons prey on hatchlings, big 'gators do, too, and catfish, and water turtles. The baby is almost too perfect, too colorful, finely enameled. Its head is up; it peers at us sideways. We turn, slide away, go on.

Finally we're there, wherever it is; a maze of shallow backcountry pools. We slow, the motor chutters to near-stop. The night noises — crickets, cone-heads, drones of beetles, 'skeeter whines, pale moth flip-flops against our lights — close in again to fill the world. We undo our seatbelts and stand as if perched on a tower, shining the night-blasters over the riffled water and the bending grass; there they are, small twin lights of little 'gators, moving, silent.

One doesn't think of reptiles as having voices, but Dave has told me more than once that I've come here at the wrong time. What I've missed by coming now, in summer, is the night roaring of spring. In spring, in mating time, alligators roar, males and females both; they bellow at each other, slap their heads on the water, roil the water to soup with their tails. They fight: males fight with males over territory and mates, females fight with females over the best nesting grounds. A nest in a poor place will be flooded out, overheated by sun, raided by coons. The females do most of the soliciting and all of the choosing of mates: if they don't like a male they let him know and that's that. Even if they like him nothing happens fast. They flirt, play, take their time. Courting couples roar at one another and meet, again and again over days and days, touching snouts, smelling each other, chuffing. They rub their chins on each other's head, slowly, rhythmically. They blow bubbles: she at him, he at her. They ride around, playfully, on each other's back. The final twining of tails is no less slow and tender than all that has gone before.

Why should the bubbles and the tenderness come as a surprise? I don't know.

Now these backcountry pools are nursery grounds. Back here in the calm there are little ones. Dozens.

Dave nods and I climb down into the bows. The young warden fishes in a sack for a metal rod with a loop of cable on the end, a loop he can tighten; he pulls at it, back and forth, getting the feel; it's a new gadget to him, too. The motor chutters, Dave points with one hand and steers with the other, and we creep up on one three-footer, an armful of 'gator, slim and sprightly as a small dog. With a quick double tail-wave it glides toward the grass and we glide after it, are over it, looming, and the warden reaches out, the loop is around its neck and is pulled tight and up and suddenly the 'gator is lashing in the air and is in. The young warden grabs it, both hands on its jaws, the width of a sandwich; I pull off the loop, the tail lashes our knees; then a rubber band goes on its muzzle and it's taken by nose and tail and plumped into a sack.

More. We swing in the pool, the light shines: there. The young warden has put the rod aside and is rolling up his sleeves; he grins and nods at me, I roll up mine. We creep up on this one and he ducks and has it, has it just behind the head with both hands, and the tail lashes air, spraying water, thumping the deck, and then its muzzle is rubber-banded, too, and it's in the sack and we're after more. My side this time. I brace my knees and bend as if I were over the rim of a well, and grab — but it's like grabbing a live electric cable, the thigh of a giant, an anaconda — he's out of my hands, scooted, gone.

"Jeez!" Dave shouts.

The young warden laughs, shakes his head, says to me, softly: "Pretend you're falling off a cliff," he says, "*grab* to save your life."

The airboat slides over the pool, nosing, sharklike; now this is lit, now this, in strobelike flashes, disconcerting; there are 'gators everywhere. The young warden gets two more by hand and then there's one on my side again and I lie over the gunwale thinking about that *cliff,* thinking too that I look a whole lot like the illustration Dave made of the dangling GIRL with my rear in the air and the rest of me hanging out over water and the 'gator under me; *just behind the head,* I think, *right,* and I grab him as if he were a child, someone's child, anyone's, sliding over an abyss. . . . I heave him in and then see I've got him by the tail instead of by the neck or armpits so that he can't bite me, and he's lunging and snapping and he's heavier than you can imagine as if he were made of steel, lashing like a fighting dog, so I put him down. I put him down between our feet. Dave shouts: "Jeez!" again and clambers down, and has the 'gator grabbed and rubber-banded in an instant.

He climbs back up, shaking his head. We're off again.

After that I expect the power and the weight and the hard slashing tail. Every time it's like grabbing a horse by the neck, a fighting ram; but the temperature of the animal is oddly neutral, like a riverbed stone, neither hot nor cool; but with the tense solidity of meat.

'Gators have a smell. It's a sour smell, musky, hard to describe; bitter, though, strong. Like nothing I've smelt before. Fresh, not bad, but strange.

And their jaws — holding them closed they're solid, all bone underneath. They remind me of ducks' beaks, but big, hard, like leather over coral. No give, none at all.

My arms ache; I could go on all night; but after an hour or so we've got all four of the gunnysacks filled. That's all the sacks we have. The young warden, disheveled, wet, as I am, laughing, raises his eyebrows.

"Let's go!" Dave says, "Come on!"

We hang on tight and we roar to speed again and head back over the dark grass, and back up over the levee and down bird bayou, still shining the lights; and eyes shine back. Sometimes all we see is eyes; sometimes we can creep up on this one and that one to watch them dive and swim and move away, lazily, smoothly, into the brown bayou.

Back at the camp we park the airboat, teetering, and carry the sacks into a screened room downstairs to get out of the 'skeeters. We go up and get cold sodas, and Dave's gear, then we empty the sacks on the floor. Now in the light we can see what we have; they're all between three and four feet long, all, Dave tells us, between three and four years old. They lie on the cool cement, safely rubber-banded but oddly calm, bemused; they turn their heads, a tail moves, nothing more. What I see, now that I can see, is difference. Each one is completely different in patterning and shape — some are dark, some pale, some boldly marked, others almost uniform in color, others smudged, some freckled, some narrow and long, others all belly; some have bulgy noses, some wider or flatter noses. Why spray difference so thoroughly in the world?

As we handle them — to tag them, weigh them, sex them — we see other things. One has a deformed tail as if the raw clay that was meant to be incised into complex patterning had been given a savage squeeze and twist. Another has a cleft palate, another an undershot bottom jaw, others have toes missing, dished faces, unhealed wounds.

One by one they're measured and sexed and tagged with metal tags. We move them and move them again; Dave does the tagging and writes the data down on a yellow pad, in neat columns.

"That one. There. No! That one! Yes!" he says.

The little 'gators are calm through all of this, they hardly budge. Scutes are cut from their tails in a particular pattern, unique for each, so that if they're caught again they can be identified even if the tag has worn off. We've got one that was caught and marked before, and Dave tells us — flipping back through pages and pages — that it's grown three inches in a single month. This is prime time for growing; summertime.

Outside the screen walls of the room the night is black; we are in with our harvest, bent over, under a single bulb; the data will reveal, someday, more of how they grow, where they grow, where they go, how long they live, how they live. What their life is, out there.

Now, what I'm gathering in is through my hands. Hauling them, touching them, I feel how they're made. The belly skin is soft like heavy silk, finest and softest around the jowls. The raised tiles are like fingernail there; smooth as enamel, tawny gold. Up over the flanks the tilings change, widen, darken, harden to armor, until the plaques on the back are large raised bosses like turtles' shells; greeny dark and the density of hoof. The saw-toothed line of scutes that runs down the back and along the tail, like the ridge of a dragon's spine, is as flexible as plastic or waxed cardboard; and tough, like glued shaved flakes of auto tires.

As we work, I think of dinosaurs, again; were they made like this? I can see them like this, smoothly tiled, all fine mosaicked pattern.

I can even see color: if the 'gator is of the swamp then the gold and moss greens here make sense; but for land saurians I can pull in a full reptilian palette: brilliant lizard blues, cool ivories, dark tigerish weavings, blush pink.

We run out of tags. Dave thumps down his pad and goes off, slamming the door, stomping up the stairs.

The young warden says:

"Don't mind Dave. He's that way. We all have to be a certain way, that's how God made us."

"I don't mind."

"Like the 'gators. God made them each different, like us."

He looks at the animals and smiles; he looks utterly happy, almost beatific.

"This is what I wanted to do," he says, softly. "I wanted to be a warden now for years. Just got my uniform two weeks ago. I just started. This is what I wanted to do."

He has his hand on his uniform shirt — wet, streaked with muck — but his face is the face of a fourteenth-century altarpiece saint. He shines.

"Those were the first legal 'gators I ever caught," he says.

Dave comes back.

"Where were we?" he says.

On we go. Soon enough we're finished; the tags are packed, the knife, the pad, the gear.

Then we take the rubber bands off the 'gators' mouths and carry them, one by one, out to the bayou and drop them in. They bob up, their eyes very bright. They swim slowly off. They disappear in pieces; ducking and rising, congealing to bits of ragged dark in the choppy night water, in which they are finally gone. Only water.

"Why don't you take them back where you found them?" I ask.

"They'll find their way home," Dave says.

"How do they do that?" I ask.

"How does anyone do that?" He frowns. "They do. Lots of things about 'gators we'll never know," he says, now softly, suddenly. "Lots of things. We'll never know the whole story. Never. You better believe it."

1:00 A.M.

Three nights and days have gone by and tomorrow I'm going, too. I don't want to leave; I'd tear up my ticket home in a flash. I'd settle in, learn to trap crabs. I'd stay, grabbing little 'gators by the neck, the armpits. I'd learn to catch mullet and baitfish and shrimp in the round nets, weighted at their fringes, that one throws like lassoes so they open out and fall and come up round, pulsing, dripping, like the nets of South American Indians. The 'gator hunters throw them, standing, out of their boats. I watch. The weights hit the night water in circles, sending up orbits of cool light: phosphorescence.

Scientist or fisherman, you never know what you'll get, either way. What's caught me is the never knowing. What's got to me is the live smell of the place in the dark, the monumental sighing of dark grass, the size of lit sky. It's got to everyone.

Of all of the people here, Johnnie Tarver has been the biggest surprise. I suppose after talking with him on the phone I thought he would be ham-fisted and backslapping but he isn't, he's quiet.

He's more subtle somehow than I'd expected an administrator to be, who has first of all to be a consummate politician, and that's what he is — a master of *Realpolitik* — but if he manipulates anyone here he does it so exquisitely that no one feels a thing. He's the boss and you can sense it when he walks in the room, everyone straightens up, but he doesn't seem to notice this and goes about his business, smiling, mostly. Whatever he says is understated. He doesn't talk as much as the others. He enjoys himself; he loves it out here, you can tell. This is what the desk stuff is all about. And he listens and listens well. This evening I discover that he even listens to what isn't said.

Tonight we've destroyed most of a huge box full of boiled crabs and now the arguments are heating up; the conversation and the bourbon drinking are raging around the table again and it's just as well. Johnnie told me months ago that I would be given data — wasn't it the job of writers, like scientists, to gather all the data they could? Well, yes, of course, that's it, I said, not knowing what he had planned. Greased by bourbon, it all comes out around the table after night falls: biology, politics, experience, theoretical principles, grouses, ideologies, tall tales, personal histories; and weaving through it all a common ground: a love of marsh. No pen, no tape recorder, no memory, could gather it in.

Sometime after midnight I escape the melee to write some of this down while it's all still fresh. I sit cross-legged on my bed, typing, and there is a knock on the door.

"It's me, Johnny."

"Come in," I say.

"I hope we didn't scare you off," he says.

"No, I just have to write it down. I forget too much."

He sits. He doesn't say anything at first. He has that way about him so that at first I think I'm going to get a talking-to; it is as if he were my boss.

"You know," he says at last, "I think I know what you're after here."

"You do?"

"I think so," he says. "I'm going to tell you a story about mullet. Do you know what mullet are?"

I nod.

"I think they were mullet. To tell you the truth, I don't know what they were."

There's a pause.

"This happened a couple of years ago," he says.

He was on a boat in the Gulf, at night, just beyond the marsh. There was no moon and the phosphorescence in the water was intense, the boat's wake a flowing path like the trail of a comet. They met a school of mullet, or something; anyway, whatever they were they took off in a pulse of light, trailing glowing streaks like trails in a massive cloud-chamber. Blinded by their own light they bumped other mullet or other somethings, which took off and bumped more which flew off again until the whole bay was a sea of streaming paths and flashes of collision. It was a bayful of explosion: of critical mass in dark water.

For the minutes that this lasted Johnnie and the people with him stood on the deck of the boat, transfixed. What was this — panic? Dance? Whatever it was to the fish, to the men it was spectacle; one of those moments that one is granted and that one doesn't forget.

"I don't know what that was," Johnny says. "I haven't told anyone about it, really. And the people who were there . . . we don't talk about it."

"Thank you," I say.

"Sure," he says. "That's what you're after, isn't it? Things like that. Where no one knows the answers?"

"Yes."

He gets up and goes to the door.

"You found some of that here?" he asks.

"Yes."

"Good night then."

"Good night."